LOGIC

LOGIC
Inquiry, Argument, and Order

SCOTT L. PRATT
University of Oregon
Department of Philosophy
Eugene, Oregon

Library of Congress Cataloging-in-Publication Data:

Pratt, Scott L.
    Logic : inquiry, argument, and order / Scott L. Pratt.
      p. cm.
    Includes bibliographical references and index.
    ISBN 978-0-470-37376-7 (cloth) ISBN 978-1-119-05099-5 (pbk)
  1. Logic. I. Title.
    BC71.P67 2009
    160–dc22

                            2009004195

10 9 8 7 6 5 4 3 2 1

For Alexander and Aaron

TABLE OF CONTENTS

ACKNOWLEDGMENTS

I would like to thank J. Brent Crouch, whose commitment to the study of the logic of Josiah Royce led to the beginning of this project; John Kaag, who served as my research assistant in the early stages; José Mendoza, who spent untold hours as my research assistant in the last stages of this project; and Aaron Pratt, who read and commented on the manuscript. I would also like to thank my colleagues John Lysaker, Mark Johnson, Erin McKenna, and Douglas Anderson, who asked key questions at crucial moments and provided ongoing encouragement. I would also like to thank the students who took Philosophy 325 at the University of Oregon and worked with successive stages of this book, providing invaluable feedback in the context of class discussion and outside class through helpful suggestions and questions that led to better presentations of the material. Finally, I would like to thank Scott Coltrane, Dean of the College of Arts and Sciences, who provided the research support necessary to complete this work.

The author and publishers gratefully acknowledge the following for permission to reprint copyright material:

W. V. Quine, "Two Dogmas of Empiricism (excerpts)," in *The Philosophical Review*, Volume 69, 20–43, Copyright 1951, Sage School of Philosophy at Cornell University. All rights reserved. Used by Permission of the present publisher, Duke University Press.

William Ockham, Epistemological Problems (excerpts), in *Medieval Philosophy: Selected Readings from Augustine to Buridan*, edited and introduced by Herman Shapiro, New York: Modern Library, 1964. Used by permission of the copyright holder, Hackett Publishing Company.

Rudolf Carnap, The Elimination of Metaphysics Through the Logical Analysis of Language, from Logical Positivism, edited by A. J. Ayer. Copyright © 1959 by The Free Press. Reprinted with permission of The Free Press, a Division of Simon & Schuster, Inc. All rights reserved.

The author and publishers apologize for any errors or omissions in the above list, and would be grateful to be notified of any corrections that should be incorporated in the next edition or reprint of this book.

THE SIGNIFICANCE OF LOGIC

Franz Fanon, in *The Wretched of the Earth*, identifies logic as the source of violence and oppression because it provides a framework for the division of the world into the colonizers or "settlers" and the colonized or "natives." "The colonial world," he says, "is a world cut in two. The dividing line, the frontiers are shown by barracks and police stations" (Fanon, 1963, 38). Speaking of the experience of the former French colonies of Northern Africa, Fanon says "The zone where the native lives is not complementary to the zone inhabited by the settlers. The two zones are opposed, but not in the service of a higher unity" (Fanon, 1963, 38). Instead, the division between zones is sharp and affords no place to stand. "Obedient to the rules of pure Aristotelian logic," he says, "both [zones] follow the principle of reciprocal exclusivity" (Fanon, 1963, 38–9). The division is not a division of equals, however. The exclusivity— settlers in one zone and natives in another—provides a framework for good and evil. While the settler zone reaps the benefits of the labor and land of the natives, the native zone "is a place of ill fame, peopled by men of evil repute. They are born there, it matters little where or how, and they die there, it matters not where or how" (Fanon, 1963, 39). Once the world is divided, "good," "white," rich settlers reign over the "evil," "dark," poor. The reason for this state of affairs, according to Fanon, is neither the superiority of one group nor the result of some preexisting natural law, but rather a product of the logic of separation itself.

Logic: Inquiry, Argument, and Order, by Scott L. Pratt
Copyright © 2010 John Wiley & Sons, Inc.

Logic is often treated as a sterile subject, removed from pressing issues like those identified by Fanon. The result of the removal is that logic becomes a "formal" study, ideally devoid of content and focused on the ways in which the structure of thought or language operates. This approach, however, is difficult to accept if, like Fanon, you become convinced that the landscape of oppression is a direct consequence of such a formal structure. One of the crucial reasons to study logic in 21st century America is not that it provides a "neutral" standpoint free from the pressures of subordination and prejudice, but because logic itself has been seen as part of the problem. This is perhaps even more important in our post-9/11 world in which stark cultural and religious differences seem to guarantee that traditional logic will be at a loss to respond. From the traditional perspective, if "their" claims are true, we conclude, ours must necessarily be false. "Our" survival then is directly dependent upon affirming "our" truths and guaranteeing that "theirs" are false. Our reasoning seems impeccable. And yet, many people suspect that a mistake has been made at some level and our conviction that "their beliefs are false" is more a hope than a certainty.

In the next chapter, we will consider what logic is in more detail. For now, it is useful to think of logic as a study of the principles that order the relations among claims about the world. To the extent that judgments and actions follow from these claims, logic also has something to say about the world as well. Fanon's challenge to logic is made in the context of an effort to change the circumstances of people subjected to colonialism. He sees the circumstances of these people as closely bound up with how claims about the world direct action, justify policies, and set expectations about the future. To challenge the structure of colonialism, Fanon believes that the system used to order claims about the world needs to be changed as well. For Fanon, the ability to change the life of the oppressed requires changes in circumstances *and* changes in the system of logic that frames how the circumstances are understood. What at first appears to be a political or moral issue turns out to be an issue of logic as well.

As we begin a broad study of logic, its history, its philosophical foundations, and its formal structures, let us first consider some of the problems that, like Fanon's concern about the logic of colonialism, draw us into the study of logic. There are four kinds of problems that both call for the study of logic and provide a set of issues that can focus our attention on particular aspects of logic. The problems are, first, *the problem of abstraction*, second, *the problem of dualism*, third, *the problem of incommensurability*, and the fourth, *the problem of boundaries*.

1.1 THE PROBLEM OF ABSTRACTION

The first problem—abstraction—is developed at length by Andrea Nye in her book, *Words of Power: A Feminist Reading of the History of Logic*. In Nye's examination, logic, as a field of study and as a way of understanding human reason, is grounded in a tradition of thinkers who divide "form" from content. Nye identifies the beginning of this tradition in the writings of Parmenides, who viewed the study of logic as a quest for harmony outside the particulars and problems of daily life. By dividing the changing particular world from the things that are unchanging, Parmenides seeks to enter the realm of logic where, says Nye, "[removed] from the uncertain pleasures of the flesh, Parmenides is reborn. ... He has left behind the life of the towns. He has left behind the community where the only antidote to the painful realization that life consists in both happiness and pain, light and darkness, good and evil is the certainty that dark works are always eventually, if temporarily, overcome by purity and light. The logician will not accept such a compromised comfort" (Nye, 1990, 19). The crux of Parmenides' logic is the claim that "Never shall this thought prevail, that not-being is; Nay, keep your mind from this path of investigation," (Plato, 1961, 258d). The realm of logic, therefore, is a realm in which things just are what they are and claims of relation, that one thing is not another, are inadmissible; such claims are the features of ambiguous lived experience. The notion of *negation*—of something being what it is by not being something else—is the central issue.

Plato challenges the Parmenidean model in *The Sophist* by establishing that negation is not a mark of nonexistence but rather a mark of difference. The Eleatic Stranger explains, "Then when we are told that the negative signifies the opposite, we shall not admit it; we shall admit only that the particle 'not' [*ou* and *mê*] indicates something different from the words to which it is prefixed, or rather from the things denoted by the words that follow the negative" (Plato, 1961, 257b–c). From this perspective, logic is no longer a study of self-identity (if that even counts as a study); it is a study of relations and so opens up recognition of a hierarchy of differences. Socrates is a human being and so different from tigers and squirrels, that is, kinds that are not human. But he is also a mammal, and an animal. To understand what Socrates is, one asks at each level whether Socrates is in the category or not (or has those characteristics or not). By this process, a thing can be classified as a certain kind of thing dependent upon the established relations among the categories or characteristics that mark differences between things.

Where Parmenidean logic took the logician away from the world of particulars, Plato established a structure independent of particulars that can be used to understand them. However, as Nye observes, the resulting conceptual or formal order, to the degree that it reflects the prejudices and hierarchies of those making the divisions, also amounts to a means "with which to impose one's view on others" (Nye, 1990, 29). Much depends on who establishes the divisions and hierarchies. Plato, in his dialogue *Cratylus*, argues through Socrates that the surface features of a thing—its appearance—are not what a philosopher seeks. Instead, the philosopher (or in this case the logician) "must be a butcher who trims the fleshy fat of ordinary talk to the bare bones of truth" (Nye, 1990, 33). It is the work of logical division to divide the world at the joints (that is, literally to "analyze" the world). Such divisions, Nye argues, makes possible conversations in which "one party [the butcher, as it were] is in complete control of the discussion" (Nye, 1990, 33). In order to participate in such a conversation, those who have been excluded from the process of dividing up the world must themselves conform to the established structures. In a similar way, for Fanon, such a requirement also means that those who live within a structure of divisions must take on the consequences of logic even to consider their own situation. "In order to assimilate and to experience the oppressor's culture," Fanon observes, "the native has had to leave certain of his intellectual possessions in pawn. These pledges include his adoption of the forms of thought of the colonialist bourgeoisie" (Fanon, 1963, 49).

The issue, again following Nye, is that logic in its earliest Western form institutes a separation from the world, adopts a structure that is in accord with the interests of a certain class and gender, and then is returned to the world as an absolute structure which necessarily frames human interactions and experience. The history of logic as a field of study, she argues, is a history of continuing to enforce the separation of formal structure and the experience of the world. The result of such formalism is in part to establish a fixed hierarchy in experience and to restrict reform to a process that uses the very tools whose use minimize change. Writing as a feminist seeking to challenge the oppression of women, Nye concludes, "if the point of logic is to frame a way of speaking in which what another says does not have to be heard or understood, in which only the voice of a unitary authority is meaningful, in which we can avoid understanding even what we say to ourselves, then no application of logic can be feminist" (Nye, 1990, 179). In short, the application of logic cannot address the status of women—or, one suspects, the status of anyone whose place is fixed within a hierarchy they cannot change. The solution to the problem is to adopt an alternative way of thinking which gives up the practice of defining by division in favor of a strategy she calls "reading," which takes up the products of logic and challenges "its pretensions to autonomy" (Nye, 1990, 176).

Is Nye correct that logic emerges as a social construct in response to particular social situations and through the use of the process of negation? Is it right that a logic of divisions can only reinforce established social attitudes such as sexism, racism, classism, and so on? If Nye is right, then the origin of logic is a key concern for anyone who wishes to question the place of women or colonized peoples.

AN ALTERNATIVE ABSTRACTION

Mark Johnson (2007) provides an alternate way of understanding abstraction. For Johnson, abstraction consists in "attending to some aspect of a continuous situation in such a way that a quality or pattern stands out as distinguished from other patterns or aspects of the situation" (Johnson, 2007, 89). In this sense, Johnson agrees with Nye that abstraction is the result of a process of selection. What makes this process problematic for Nye is that acts of selection exert power over the things (and people) they select. In order to undermine oppression, one must avoid such exercises of power and seek other ways of engaging one's community and environment. From this angle, logic, as a discipline, is a process of reinforcing the false necessity that the only way to engage others is through the use of power. Johnson concludes, on the contrary, "Abstraction is a great tool for the furtherance of human inquiry." Yet, he continues, despite its value, "it is also responsible for much of the loss of meaning that is available to us in any given situation" (2007, 270).

The practice of abstraction leads to the formation of habits that help frame the process of interaction with others. Once established, such habits also have the potential to block aspects of an experienced situation and the result is a loss of meaning. "We tend," Johnson argues, "to abstract structures from situations and then attempt to apply them to situations that seem to us to be similar in kind to former ones. ... The mistake, one that is deeply etched in our intellectual tradition, is to think that just because certain logical relations 'transcend' particular instantiations by applying to many different situations, it would be possible to erase the embeddedness of thought in concrete, actual situations and their concrete qualities" (2007, 104). From Nye's perspective, these habits of thought can be seen as a kind of power developed in practice over time and applied in a way that controls the production of meaning. But where Nye wants to seek a process of meaning that minimizes selection, Johnson argues that human beings are naturally beings who select in order to survive and flourish. Abstraction is a necessary human practice and logic is the study of these practices. Like other practices, abstraction can be done in ways that are better and worse. Logic can, from this perspective, contribute to wider projects of liberation. The central implication of Johnson's argument is that "If logic doesn't merely fall down from the Platonic heavens above, then it must surely rise up from our embodied experience as functioning organisms within changing environments" (2007, 102) and this makes it subject to criticism and change.

1.2 THE PROBLEM OF DUALISM

Philosopher Val Plumwood challenges Nye's diagnosis of logic's problems in her paper "The Politics of Reason: Towards a Feminist Logic." From her perspective, Nye's condemnation of the logical process of negation conflates two distinct issues: division (or dichotomy) and domination (or dualism). Plumwood argues that Nye's critique of logic overlooks the importance of the operations of division for thinking about the world. Equating dualism and dichotomy, she continues, "would either cripple all thought (if we were forced to abandon dichotomy along with dualism) or collapse the concept of dualism (if we were forced to retain dualism along with dichotomy). In either case, escape from dualism becomes impossible" (Plumwood, 2002, 22). It is not negation (or the process of division), but rather the way that negation has been "treated" that has led to the problems identified by Nye and, by extension, Fanon. *Dualism*, Plumwood says, "can be seen as an alienated form of differentiation, in which power construes and constructs difference in terms of an inferior and alien realm" (Plumwood, 2002, 19). *Dichotomy* (literally, "cutting in two"), on the other hand, is "simply making a division or drawing a distinction" (Plumwood, 2002, 22). Dualism is "more than a relation of dichotomy, difference or non-identity, and more than a simple hierarchical relationship." Instead, dualism is a complex multifaceted relation that begins with division but becomes "a relation of separation and domination inscribed and naturalized in culture and characterized by radical exclusion, distancing and opposition between orders construed as systematically higher and lower, as inferior and superior, ruler and ruled, center and periphery" (2002, 23). Dualisms, Plumwood argues, "are not universal features of human thought, but conceptual resources to and foundations for social domination" (2002, 20). Dualisms relevant to Western culture include "the contrasting pairs of culture/ nature, reason/nature, male/female, mind/body, master/slave, reason/matter (physicality), rationality/animality, reason/emotion, mind (spirit)/nature, civilized/primitive (nature), production/reproduction (nature), public/private, subject/object, self/other" (Plumwood, 2002, 30).

Plumwood identifies five features of dualism that illustrate its potential to dominate. The first feature, *backgrounding*, takes the division, for example, between male and female, and makes one side of the dualism "inessential" background for the other. The important side of the dualism becomes the source of value for the relation: women literally stand in the background, unrelated to the valued activities of men. The second feature, *radical exclusion*, takes the division between the terms of the dualism as marking an absolute difference. If races, for example, are divided into white and non-white they have no essential features in common and must be understood as fundamentally distinct. Claims about commonalities that support equal treatment or opportunity are necessarily suspect since there is no reason to apply principles relevant to one side of the dualism to the other side as well. If race differences are radically exclusive, then the idea of a common humanity is either a mistake or is not an essential characteristic of either side.

The third feature of dualism, *relational definition*, argues that one side of the dualism defines the other. Consider the dualism reason and emotion. In this case, whatever is essential to defining reason serves to define emotion as well. If reason is a process or attitude that is "objective" and formal, then emotion is characterized by the lack of these characteristics—it is "subjective" and informal (or unstructured). The fourth feature Plumwood associates with dualism is *instrumentalism*, in which one side of a dualism is identified as being in service to the other. The culture/nature divide generally accompanies the idea that nature is to be a benefit to human beings who have dominion over it. Nature, in this sense, is a resource for culture even as it is divided from it and "essentially" different. The final feature, *homogenization*, holds that things that are part of one side of a dualism must be seen as "the same" as each other and so different from all things on the other side. It is important for the integrity of the division to find that all men are the same in all the ways that they differ from women and vice versa. Apparent differences among men—especially if these are differences that suggest similarities with women— are not allowed because they would undermine the dualism.

Plumwood argues that by understanding the notion of dichotomy and the ways that division operates within a cultural context, we can begin to think about alternative logics. What are the implications of rejecting dualism and accepting the notion of division as our logical starting point? Will this provide a means for rethinking the ways in which we accept and reject claims about the world? If it does, does the notion of dichotomy provide a means of comparison among diverse claims and "worldviews" or does it assure us that comparison, at best, will be relative to our own interests? Does giving up dualism lead to some radical form of relativism?

THE GREAT CHAIN OF BEING

The Great Chain of Being, though generally taken as a metaphysical concept, provides an example of the intersection between difference and hierarchy that emerges in a dualism of the sort challenged by Plumwood. Arthur Lovejoy (1936) defined the chain of being as "a conception of the universe ... composed of an immense, or ... of an infinite, number of links ranging in hierarchical order from the meagerest kind of existents, which barely escape non-existence, through 'every possible' grade up to the *ens perfectissimum* [perfect being] ... every one of them differing from that immediately above and immediately below it by the 'least possible' degree of difference" (1936, 59). The concept of the chain of being, Lovejoy argues, emerged in the history of Western philosophy as a result of three distinct commitments that, in the course of history, interacted with each other and with the development of Christianity. The result was what might be called a habit of thought that organized the objects of experience in a way that required a hierarchical relationship among distinct things.

The three commitments were, first, what Lovejoy called the principle of plenitude. Grounded in the account of creation given in the *Timaeus*, Plato proposes that "nothing incomplete is beautiful" and that since the Good must be beautiful it must also be complete (1936, 52). In the tradition that inherited Plato, the conclusion became the principle "that no genuine potentiality of being can remain unfilled, that the extent and abundance of creation must be as great as the possibility of existence and commensurate with the productive capacity of a 'perfect' and inexhaustible Source, and that the world is the better, the more things it contains" (1936, 52).

Aristotle responded to Plato's demand for a full universe with the recognition that the universe must also be continuous—the second formative commitment. "Things are said to be continuous," Aristotle explained, "whenever there is one and the same limit of both wherein they overlap and which they possess in common" (quoted in Lovejoy, 1936, 55). On one hand, continuity helped to ensure that the universe is as full as it must be in order to be Good, since without continuity the universe would contain gaps. On the other hand, continuity ensured that nothing could stand outside the universe because, if something did, then the universe would not be complete and so could not be Good.

The third commitment identified by Lovejoy follows from the notion of *ens perfectissimum*. This principle, which he called "unilinear gradation," recognized that since perfection was a matter of completeness, anything less could be understood in terms of privation, that is, incompleteness. Lovejoy quotes Aristotle: "all individual things may be graded according to the degree to which they are infected with [mere] potentiality" (1936, 59). The resulting expectation is that given any two things, to the extent that they are individuals (something with boundaries), they will lack something and since they are distinct individuals, they will lack in different ways. "Difference of kind," Lovejoy concludes, "is treated as necessarily equivalent to difference of excellence, to diversity of rank in a hierarchy" (1936, 64).

While the chain of being included everything (and so was equivalent to the universe as a whole), it also established that, for any two kinds of things, they would differ in the degree to which they have or lack excellence. The metaphysical principle is also a logical one because it establishes a formal framework in terms of which things can be understood. In this context, no dichotomy can ever be just a dichotomy. It must also necessarily include a relationship of relative value given in the nature of things. In the end, Lovejoy argues that the chain of being lost its force as a formative principle in the 19th century with the advent of evolution. "[T]he history of the idea of the Chain of Being … is the history of a failure; more precisely and more justly, it is the record of an experiment in thought carried on for many centuries by many great and lesser minds, which can now be seen to have had an instructive negative outcome" (1936, 329).

Plumwood would certainly agree that the idea led to a negative outcome, but would probably argue that the habit of thought that instituted dualisms in the context of the chain of being did not fade away or transform, but rather continues to operate in the 21st century.

1.3 THE PROBLEM OF INCOMMENSURABILITY

The idea that division is necessary leads to a third problem. If there are divisions (but not dualistic relations) between cultures, genders, races, or even sciences, one wonders whether the differences make communication across these differences impossible. This is the problem of *incommensurability*. Thomas Kuhn, in his *The Structure of Scientific Revolutions*, develops the classic formulation of this problem when he proposes that science is framed by paradigms that determine scientists' expectations and ways of interpreting data. When claims are made within one paradigm, they literally do not make sense in another. This failure of understanding takes several forms. When competing paradigms emerge, scientists within one paradigm have difficulty agreeing with scientists within another on what problems need to be addressed. The resulting disagreement serves as a ground for dismissing the views of those who do not share one's paradigm. "[The] most fundamental aspect of the incommensurability of competing paradigms," he says, is that "the proponents of competing paradigms practice their trades in different worlds" (Kuhn, 1970, 150). Some communication between paradigms is possible by means of translation, Kuhn claims, but the communication is partial at best. To make the alternative *weltanschauung* or worldview one's own, one must "go native, discover that one is thinking and working in, not simply translating out of, a language that was previously foreign" (Kuhn, 1970, 204). Translation does not solve the problem but "going native," since it leaves the old paradigm behind, does not solve it either.

Others, looking at the problem of incommensurability, have argued that the problem is not a real one at all. Donald Davidson, for example, argues that while there are practical problems of understanding among people who speak different languages, these are only problems of translation and not the apparently insurmountable problem of incommensurability.[1] For Davidson, that we can even identify problems of translation means that, at some level, we and speakers of the other language share a basic paradigm or "conceptual scheme" in terms of which translation can begin. In this way, incommensurability is not a real problem for logic. In fact, logic provides the common ground of reasoning principles that we use to overcome such difficulties. Despite Davidson's conclusions, however, his response is persuasive in only some circumstances; there are others in which it appears that logic is useless in undercutting apparently irresolvable differences.

Consider the comparison of Islam and "modernity" proposed by Seyyed Hossein Nasr. For Nasr, the "world of Islam" and the "modern world" are sharply separated. At the center of the difference are contrasting conceptions of human nature. Islam, Nasr argues, adopts the conception of "*homo islamicus*" in which human beings are properly understood as "the slave[s] of God (*al-'abd*) and His vice-regent[s] on earth." As such, *homo islamicus* "rules over the earth," but, as God's vice-regents, also "bears responsibility for the created order before the Creator." To carry out this role, human beings possess reason but also "the possibility of inward knowledge, the knowledge of his own inner being which is key to the knowledge of God." In contrast, Nasr holds that the "modern" conception of a human being is "an animal which happens to speak and think," who is "purely an earthly creature, master of nature, responsible to no one but himself." These contrasting conceptions of human nature, Nasr argues, cannot be "harmonized."

This failure to harmonize is well illustrated by claims about the origins of life. The theory of evolution, a view compatible with the notion of human nature held by "modernism," holds that human culture is a matter of chance, a process without purpose or end. In this paradigm, human activities are properly understood as a product of a particular process of chance and selection and humanity's autonomy is a consequence of the lack of an end or purpose other than human purpose. The Islamic conception—whether one accepts some form of biological development or not—nevertheless recognizes the role of purpose and the context of "higher states of being and the archetypal realities which determine the forms of this world."[2] Even if the opposed concepts can be identified through translation (as they clearly can be), the meaning and import of the conflicting claims cannot be resolved unless one gives up one paradigm for the other.[3]

---

[1] See Davidson, 2001.

[2] All quotations of Nasr are from Nasr, 2003.

[3] The two paragraphs above are taken from Pratt, 2007, 107.

How can such a conflict in views be reconciled? Is this—as some would have it—a matter of conflicting values as distinct from a "logical" problem? That there is a conflict demonstrates that both "modernists" and Muslims share some common ground, enough that they can conclude that their views are incompatible, that is, that their views cannot both be true. The conflict also confirms that there is something more at stake because there appears to be no way to resolve one set of conclusions in terms of the other. This becomes a logical matter if we see that the ability to compare views also requires some system of order in terms of which a comparison can go on. The study of logic provides a context for seeking this means of comparison.

ANOTHER FORM OF INCOMMENSURABILITY

Ofelia Schutte (2000) takes up incommensurability in the context of cross-cultural communication when "one culture circumstantially holds the upper hand over another" (2000, 50). In such cases, there is "a residue of meaning that will not be reached," which can be understood either as a kind of "arithmetic incommensurability" or as a "strangeness" that marks a "displacement of usual expectation." The former understanding at once recognizes "on the plus side" the possibility of a "conveyable" message and "on the minus side" that whatever meaning is culturally distinctive will not be conveyed. In order to improve cross-cultural communication, participants can try to maximize the meaning that can be shared and minimize the meaning that is bound to a speaker's cultural context. The uncommunicated remainder may or may not be problematic in a particular communication.

The alternative understanding is that incommensurability marks something strange that is always present in the context of cross-cultural communication. To fail to bypass the strangeness without acknowledgement or "to subsume them under an already familiar category" is to override the cross-cultural situation. Since the ability to override the situation is most often one possessed by those of the dominant culture, cross-cultural communication becomes an instance of exclusion and oppression in which what makes a person distinctive is ignored and whatever contribution one could make from a cultural perspective is lost.

To understand this process framed by incommensurability, Schutte asks what it means to be culturally different, that is, not of the dominant culture. As a Latina, she observes, she is cast "in a recognizable category, through which the meaning of [her] difference is delimited according to whatever set of associations this term may evoke" (2000, 52). Taken together, these associations mark her as invisible "as a producer of culture." Latinas, she concludes, "are viewed principally as caregivers whose function in culture is to transmit and conserve, not question and create, cultural values" (2000, 53). Schutte observes that it is possible for her to find a place in a conversation with members of the dominant culture but only by knowing and practicing "the language and epistemic maneuvers" of that culture. The result is that she becomes a "split subject" such that, when she acts as "herself," she is viewed as strange; when she performs the "speaking position expected of the dominant culture, [she] is recognized as a real agent in the real world" (2000, 54).

For Schutte, incommensurability does not mark completely disconnected paradigms, but rather a practice in which something is experienced in common (the context of communication, in this case) and something else is not communicable and is experienced as strangeness. When Schutte encounters a colleague whose speaker position is Anglophone, he does not speak to her, but to her-as-a-non-Latina and in so doing misses part of the message she gives in reply. If the interlocutor recognized that he was missing something, he has the potential both to see that she is different from him and, in doing so, to recognize that there are speaker locations outside the dominant culture. This "radical recentering" would end up making "demands in his everyday world. ... He might have to acknowledge his own split subjectivity, change his fixed way of life, welcome the stranger within, and perhaps alter his views and relations with others in ways he had not foreseen" (2000, 54).

Schutte traces the process of failed cross-cultural communication to philosophy. "Philosophers," she says, "are often taught that philosophical claims can be stated in a language that is essentially outside of culture" (2000, 55). When such philosophers are willing to acknowledge the presence of something incommensurable in the context of cross-cultural communication, "The debate lies in whether such incommensurable elements should be assigned to what is irrelevant to philosophical meaning and knowledge, and thus irrelevant to the operations of reason; or whether ... the incommensurable elements should be seen as inherent to the processes of reasoning itself" (2000, 55). Schutte argues for the latter conclusion on the grounds that communication requires both that participants understand what is being said and be able to "relate what is being said to a complex of signifiers, denoting or somehow pointing to what remains unsaid" (2000, 55–6). Communication that leads to exclusion of culturally different others is, in part, a failure on the part of the dominant participants to recognize the message conveyed—both in what is said and in what is unsaid. The underlying process of exclusion marks the operation of a system of order that frames the resulting "partial" incommensurability.

1.4   THE PROBLEM OF BOUNDARIES

The idea that we can somehow seek a means of comparison recalls the problem of how we are to understand the process of division. On one hand, if division requires a relation of dependence, then we seem to risk the relation of dualism. From this angle, we can compare claims, but can this be done without reinstituting the problematic hierarchies identified by Plumwood? If we reject comparative relations, however, we seem to face a situation in which diverse views are inaccessible to one another and logic becomes a parochial matter, disconnected again from the problems faced by people in a complex world. In this case, comparison cannot be resolved across different worlds, as Kuhn concludes, "A decision of that kind can only be made on faith" (Kuhn, 1970, 158). This tension—between dualism and incommensurability—leads to a fourth problem that can motivate us to the study of logic: the problem of boundaries.

The problem is most clearly, though somewhat abstractly, seen in an example proposed by C. S. Peirce. The issue in question is how we understand a division established to define two regions of an "original vague potentiality."

The problem is comparable to any effort to institute a definition in terms of which things can be identified as particular things or kinds of things. Peirce asks that we begin with a blackboard which serves as a two-dimensional continuum, the region of vague potentiality. "I draw a chalk line," he says, introducing a discontinuity to the region which he describes as "one of those brute acts by which alone the original vagueness could have made a step toward definiteness." The line he has drawn is, he observes, not a series of points, but an oval, a two-dimensional figure introduced in relation to the otherwise black field. The white chalk line is "something new," but "the boundary between the black and white is neither black, nor white, nor neither, nor both. It is the pairedness of the two" (6.203). The *boundary* is some *thing*, but its logical status is ambiguous. Despite its ambiguity, however, the boundary remains necessary if there are to be both black and white.[4]

The same logical relation seems to emerge in the concerns raised by the idea of personal identity and its implications for human society. In his 1903 collection of essays, *The Souls of Black Folk*, W. E. B. Du Bois framed the problem of boundaries as a matter of "double consciousness."

> After the Egyptian and Indian, the Greek and Roman, the Teuton and Mongolian, the Negro is a sort of seventh son, born with a veil, and gifted with second-sight in this American world,—a world which yields him no true self-consciousness, but only lets him see himself through the revelation of the other world. It is a peculiar sensation, this double-consciousness, this sense of always looking at one's self through the eyes of others, of measuring one's soul by the tape of a world that looks on in amused contempt and pity. One ever feels his twoness,— an American, a Negro; two souls, two thoughts, two unreconciled strivings; two warring ideals in one dark body, whose dogged strength alone keeps it from being torn asunder. (Du Bois, 1903, 3)

Faced with a social and historical situation that both demanded and expected identification with the American nation and at the same time which denied him access to full participation, Du Bois came to see himself of both sides and of neither, a new kind of person at the boundary between cultures. By inheriting or at least finding himself in a context defined by a logic of inclusion and exclusion, Du Bois also found himself almost literally a living contradiction. From this painful situation, however, Du Bois also was empowered as an activist who, as he said later in his essay, "The Souls of White Folk," could see and know things that could not be seen or known by those who stood firmly within "white" America (Du Bois, 1920, 29). The ambiguity and tension of the boundary, however, took its toll. Late in his life (in 1963), as if at last to concede to the "logic" that conditioned his identity, he renounced his United States citizenship and went to Ghana—a kind of return, he said, to the home of his ancestors, but one that seemed to adopt the logic of exclusion and set aside his "twoness."

---

[4] All quotations from Peirce, except where noted, are from Peirce, 1994, and designated by the standard four-digit section number, e.g., 6.203.

In 1993, Gloria Anzaldúa identified a similar boundary experience in her book, *Borderlands/La Frontera*. She introduces the experience in a poem "To live in the Borderlands means you": "are neither *hispana India negra española/ni gabacha, eres mestiza, mulata*, half-breed/caught in the crossfire between camps/while carrying all five races on your back/not knowing which side to turn to, run from ..." (Anzaldúa, 1999, 216). For Anzaldúa, the experience of living between the political nations is not a matter of simple identification with one side or the other, but rather a matter of the creative emergence of something born to both sides and yet distinct from both. This new person, she argues, is a *mestiza* consciousness. *Mestizas* are between political nations, the U.S. and Mexico; between cultures, the indigenous culture of the Aztecs and the inherited culture of the Spanish colonizers; between the rich and poor; between men and women. In the presence of each of these dualisms, Anzaldúa argues that she and others like her are of both sides and of neither at once.

Just as Du Bois argued that this "between" perspective leads to a distinct epistemic position, Anzaldúa sees *mestiza* people as distinctive sources of transformation. She writes "The work of mestiza consciousness is to break down the subject-object duality that keeps her prisoner and to show in the flesh through the images in her work how duality is transcended. ... A massive uprooting of dualistic thinking in the individual and collective consciousness is the beginning of a long struggle, but one that could, in our best hopes, bring us to the end of rape, of violence, of war" (Anzaldúa, 1999, 102). Such people suffer for their place and for the transformation of the communities they are bound to. They also are literally necessary for the possibility of change. "In attempting to work out a synthesis the self has added a third element which is greater than the sum of its parts. That third element is a new consciousness—a *mestiza* consciousness—and though it is a source of intense pain, its energy comes from continual creative motion that keeps breaking down the unitary aspect of each new paradigm" (Anzaldúa, 1999, 101–2).

How are we to understand the ambiguous space and people that stand between the sides of a dichotomy? To what extent do they make a "cut" or division possible? To what extent do they make the connection between the sides at once a common ground *and* incommensurable? In the following series of discussions, we will reopen a consideration of logic in an effort to understand its philosophical foundations and the emergence of its formal systems, and, as a means to understand the lived experiences of differences and similarities, of variety and common ground.

CONTINUITY

The idea of continuity marks the connections between things (broadly taken). John Dewey used the idea at once to argue against dualisms of the sort criticized by both

Nye and Plumwood and to serve as a "postulate" for his "naturalistic" approach to philosophy. The postulate of continuity applies to the "continuity of the lower (less complex) and the higher (more complex) activities and forms." Its meaning, he continues, "excludes complete rupture on one side and mere repetition of identities on the other; it precludes reduction of the 'higher' to the 'lower' just as it precludes complete breaks and gaps" (1938, 30). This principle leads to a form of "naturalism" because "What *is* excluded by the postulate … is the appearance upon the scene of a totally new outside force as a cause of changes that occur" (1938, 31). This same rejection of a break between "higher" and "lower" applies to hierarchical dualisms such as theory/practice, mind/body, and even good/evil. Given this starting place, Dewey could set aside questions of transcendence and theology and focus on the character of the interactions between organisms (including humans) and their environments as a source of the principles that order the interactions and as a source of values (including truth values). He could also challenge the separation of mind and body and the dismissal of practice by those who are interested in theory and vice versa.

Despite the appeal of the postulate and Dewey's challenges to dualism, the notion of continuity remains difficult to explain. On one hand, it serves as an obvious principle of unification, connecting things into the wholes of experience. At the same time, continuity seems to be a principle of a particular kind of division. One way in which a line is continuous, for example, is that there are no gaps between its limits. So if there is a line running from point *a* to point *b*, one should find, between those points, another point, *c*. Between *a* and the point *c*, one will find another point and so on. As a result, if the line is continuous, then it is composed of infinitely many points in a way that both marks divisions and ensures the whole has no gaps.

When C. S. Peirce examined the notion of continuity, he argued that the idea could be understood in terms of two alternate conceptions. The first, he called "Kanticity," is the idea that continuity means infinite divisibility. The second, called "Aristotlicity," captures the idea that things which are continuous are also, in some sense, complete in themselves. Aristotle makes his point by observing that two things are continuous when they share the same limits. "I call two things continuous," Aristotle says, "when the limits of each, with which they touch and by which they are kept together, become one and the same, so that plainly the continuous is found in the things out of which a unity naturally arises in virtue of their contact" (Aristotle, 1941, 1069b 5–10). Given the line *ab* just mentioned, it is continuous with itself and every other line that also has the same limits, points *a* and *b*. As Peirce explains, "a continuum contains the end point belonging to every endless series of points which it contains. An obvious corollary is that every continuum contains its limits" (6.123).

In his *Encyclopedia Logic*, Hegel presents both kinds of continuity as characteristics of quantifiable things (e.g., space, time and matter). The first (infinite divisibility) he called continuity and the second (things containing their own limits) discreteness. There is, he concluded, "no such thing as a merely Continuous or a merely Discrete quantity. We may speak of the two as two particular and opposite species of magnitude; but that is merely the result of our abstracting reflection, which in viewing definite magnitudes waives now the one, now the other, of the elements contained in inseparable unity in the notion of quantity. Thus, it may be said, the space occupied by this room is a continuous magnitude, and the hundred men assembled in it form a discrete magnitude. And yet the space is continuous and discrete at the same time" (1975, 148–9).

Later in his career Peirce argued that the two kinds of continuity were insufficient to capture the meaning of the term. The problem is a consequence of "a misunderstanding of Kant's definition which *he himself* likewise fell into." Rather than being a matter of infinite indivisibility, Peirce concludes, "Kant's real definition implies that a continuous line contains no points" (6.168). In his 1898 lectures, published as *Reasoning and the Logic of Things*, Peirce illustrates his conclusion by considering "a line that returns to itself" such as a circle. The line is "a collection of points" where "no point in this line has any distinct identity absolutely discriminated from any other." Introduce a break in the line, however, and the points that mark the break gain a distinct identity. "If there were not discontinuity," he concludes, "there would be no distinct point there,—that is, no point absolutely distinct in its being from all others" (1992, 159–60). The principle of continuity, then, leads back to the problem of boundaries.

1.5 EXAMPLES FOR DISCUSSION

Each of the following examples represents an instance of at least one of the four problems discussed above. In each case, consider the claims made by the authors, the issues raised by them and the ways in which these illustrate one of the problems of logic.

1. "There is ... intellectual virtue, which is that of generality or impartiality. I recommend the following exercise: When, in a sentence expressing political opinion, there are words that arouse powerful but different emotions in different readers, try replacing them by symbols, A, B, C, and so on, and forgetting the particular significance of the symbols. Suppose A is England, B is Germany, and C is Russia. So long as you remember what the letters mean, most of the things you will believe will depend upon whether you are English, German or Russian, which is logically irrelevant. When, in elementary algebra, you do problems about A, B, and C going up a mountain, you have no emotional interest in the gentlemen concerned, and you do your best to work out the solution with impersonal correctness. But if you thought that A was yourself, B your hated rival, and C the schoolmaster who set the problem, your calculations would go askew, and you would be sure to find that A was first and C was last. In thinking about political problems this kind of emotional bias is bound to be present, and only care and practice can enable you to think as objectively as you do in the algebraic problem." (Russell, 1950, 31)

2. "The white world of abstract symbols became a nightmare for Indian people. The words of the treaties, clearly stating that Indians should have 'free and undisturbed' use of their lands under the protection of the federal government, were cast aside by the whites as if they didn't exist. The Sioux once had a treaty plainly stating that it would take the signatures or marks of three-fourths of the adult males to amend it. Yet through force the government obtained only 10 percent of the required signatures and declared the new agreement valid. ... Words and situations never seemed to fit together. Always, it seemed, the white man chose a course of

action that did not work. The white man preached that it was good to help the poor, yet he did nothing to assist the poor in his society. Instead he put constant pressure on the Indian people to hoard their worldly goods, and when they failed to accumulate capital but freely gave to the poor the white man reacted violently. The failure of communication created a void into which poured the white do-gooder, the missionary, the promoter, the scholar, and every conceivable type of person who believed he could help. White society failed to understand the situation because this conglomerate of assistance blurred the real issue beyond recognition. ... Abstract theories create abstract action. Lumping together the variety of tribal problems and seeking the demonic principle at work which is destroying Indian people may be intellectually satisfying. But it does not change the real situation." (Deloria, 1969, 9–10, 86)

3. "The affirmative enunciation is prior to the negative for three reasons ... With respect to vocal sound, affirmative enunciation is prior to negative because it is simpler, for the negative enunciation adds a negative particle to the affirmative. With respect to thought, the affirmative enunciation, which signifies composition by the intellect, is prior to the negative, which signifies division. ... With respect to the thing, the affirmative enunciation, which signifies *to be*, is prior to the nega-tive, which signifies *not-to be*, as the having of something is naturally prior to the privation of it." (Aquinas, 1962, 64)

4. "To focus on dominance as the relevant factor in understanding gender relations is to assent to a minimalist model of society; a model that can be represented by the statement A > B ['A is greater than B']. Such an approach is not without utility as a means of focusing on a limited aspect of a society, but it is nondynamic. It may provide a static picture of an aspect of power within a society, but it is not capable of dealing with power as a process and is an insufficient base from which to analyze the construction, operation, or negotiation of gender relations within a society. The model of society implied by a complementarity model is quite different and may be represented by the statement A + B + ... N = 1. This approach, aside from being better suited to handle nonbinary gender situations, is inherently dynamic." (Sharp, 1995, 48)

5. "Since remote antiquity most people have seen one or another heavy body swing-ing back and forth on a string or chain until it finally comes to rest. To the Aristotelians, who believed that a heavy body is moved by its own nature from a higher position to a state of natural rest at a lower one, the swinging body was simply falling with difficulty. Constrained by the chain, it could achieve rest at its low point only after a tortuous motion and a considerable time. Galileo, on the other hand, looking at the swinging body, saw a pendulum, a body that almost succeeded in repeating the same motion over and over again ad infinitum. And having seen that much, Galileo observed other properties of the pendulum as well and constructed many of the most significant and original parts of his new dynamics around them. From the properties of the pendulum, for example, Galileo derived his only full and sound arguments for the independence of weight and rate of fall, as well as for the relationship between vertical height and terminal velocity of motions down inclined planes. All these natural phenomena he saw differently from the way they had been seen before." (Kuhn, 1970, 118–19)

6. "Cosmologies are the deep-rooted, symbolically expressed understandings of 'humanness.' They predate all other human structured expressions including reli-

gion and social and political orders. The first Indigenous cosmologies were based on the perception that the spirit of the universe resided in the Earth and things of the Earth, including human beings. Because of this perception, these people remained equally open to all possibilities that might manifest through the natural world. In turn, perceptions of the cycles of nature, behavior of animals, growth of plants, and interdependence of all things in nature determined culture, that is, ethics, morals, religious expression, politics, and economics. In short, they came to know and express 'natural democracy.'

"In the inclusive view of natural democracy, humans are related and interdependent with plants, animals, stones, water, clouds, and everything else. Thus, it becomes in every sense abnormal to view the world as dead matter, private property, commodities, or commercial resources. The manifestations and roots of the Native sense of democracy run much deeper than the modern American political version of democracy today in that all of nature, not only humans, has rights. This is the essential 'cosmological clash' between the foundations of Native culture and those of modern society." (Cajete, 2000, 52–3)

7. "Latin American racial formations are defined by *mestizaje*, multiracial group inclusion, and color differentiation. Racial hegemony is maintained through co-optive incorporation and reinterpretation. U.S. racial formations are defined by hypodescent, strict phenotypic differentiation that polarizes into unreal biracial categories. Racial domination is maintained through exclusion, marginalization, and repression. Today, however, due to processes of globalization, these differences in racial formations in the Americas are disappearing, even though they informed the dynamics of ethnic struggles over the last century." (Mendieta, 2000, 56)

8. "For Immanuel Kant, author of the *Critique of Practical Reason*, the absolute criterion of practical goodness cannot include any empirical, concrete content whatever. 'Good' denotes the quality of the act that can be elevated to universality—that may ethically be performed by all persons finding themselves in the same circumstances. But the universality of this action depends on the judgment of the very agent who is to perform the praxis. Thus the way is open for a surreptitious elevation of the subject's (European or capitalistic) particularity to the status of universality (validity for every culture and system). With all the 'good will' in the world, this subject can perform an objectively perverse action." (Dussel, 1988, 75–76)

9. "The totality of our so-called knowledge or beliefs, from the most casual matters of geography and history to the profoundest laws of atomic physics or even of pure mathematics and logic, is a man-made fabric which impinges on experience only along the edges. Or, to change the figure, total science is like a field of force whose boundary conditions are experience. A conflict with experience at the periphery occasions readjustments in the interior of the field. Truth values have to be redistributed over some of our statements. Re-evaluation of some statements entails re-evaluation of others, because of their logical interconnections—the logical laws being in turn simply certain further statements of the system, certain further elements of the field. Having re-evaluated one statement we must re-evaluate some others, whether they be statements logically connected with the first or whether they be the statements of logical connections themselves. But the total field is so undetermined by its boundary conditions, experience, that there is much latitude of choice as to what statements to re-evaluate in the light of any single contrary

experience. No particular experiences are linked with any particular statements in the interior of the field, except indirectly through considerations of equilibrium affecting the field as a whole.

"If this view is right, it is misleading to speak of the empirical content of an individual statement—especially if it be a statement at all remote from the experiential periphery of the field. Furthermore it becomes folly to seek a boundary between synthetic statements, which hold contingently on experience, and analytic statements which hold come what may. Any statement can be held true come what may, if we make drastic enough adjustments elsewhere in the system. Even a statement very close to the periphery can be held true in the face of recalcitrant experience by pleading hallucination or by amending certain statements of the kind called logical laws. Conversely, by the same token, no statement is immune to revision." (Quine, 1980, 42–43)

10. "Core-periphery was an essential contribution of Third World scholars. True, there had been some German geographers in the 1920s who had suggested something similar, as had Romanian sociologists in the 1930s (but the Romanians had a social structure similar to that of the Third World). But it was only when Raul Prebisch and his Latin American 'young turks' at the ECLA (Economic Commission for Latin America) got to work in the 1950s that the theme became a significant focus of social science scholarship. The basic idea was very simple. International trade was not, they said, a trade between equals. Some countries were stronger economically than others (the core) and were therefore able to trade on terms that allowed surplus-value to flow from weaker countries (the periphery) to the core. Some would later label this process 'unequal exchange'. This analysis implied a remedy for the inequality: actions by the states in the periphery to institute mechanisms that would equalize the exchange over the middle run." (Wallerstein, 2004, 11–12)

11. "The penchant of developmental theorists to project masculine image, and one that appears frightening to women, goes back at least to Freud, who built his theory of psychosexual development around the experiences of the male child that culminate in the Oedipus complex. In the 1920's, Freud struggled to resolve the contradictions posed for his theory by the differences in female anatomy and the different configuration of the young girl's early family relationships. After trying to fit women into his masculine conception, seeing them as envying that which they missed, he came instead to acknowledge, in the strength and persistence of women's pre-Oedipal attachments to their mothers, a developmental difference. He considered this difference in women's development to be responsible for what he saw as women's developmental failure." (Gilligan, 1993, 6–7)

12. "In the United States both scholars and the general public have been conditioned to view human races as natural and separate divisions within the human species based on visible physical differences. With the vast expansion of scientific knowledge in this century, however, it has become clear that human populations are not unambiguous, clearly demarcated, biologically distinct groups. Evidence from the analysis of genetics (e.g., DNA) indicates that most physical variation, about 94%, lies within so-called racial groups. Conventional geographic 'racial' groupings differ from one another only in about 6% of their genes. This means that there is greater variation within 'racial' groups than between them. In neighboring populations there is much overlapping of genes and their phenotypic (physical) expressions. Throughout history whenever different groups have come into

contact, they have interbred. The continued sharing of genetic materials has maintained all of humankind as a single species." (American Anthropological Association, 1998)

13. "Received opinion holds that racial identity is a good thing to have: Knowledge about one's 'own people' enriches an individual life, it is good to know *what* one is. This racial identity is usually based on the past, and family history is a normal starting point. However the attempt to base mixed-race identity on family history more likely than not will fail against the realities of the American biracial system. A biracial society does not, in general, support the formation of mixed-race families. And in any particular case, the facts of a mixed-race lineage may be associated with personal trauma and tragedy. Furthermore, individuals who are the first 'issues' of mixed-race in their families do not have any forebears of mixed race with whom to identify.

"The difficulties with family-history foundations for mixed-race identity lead to the question of whether an identity of mixed-race can be based on the impersonal history of mixed-race in the United States." (Zack, 1993, 69)

14. "In 'Definitional Dilemmas', Julie Greenberg tells us that legal institutions have the power to assign individuals to a particular racial or sexual category: 'Sex is still presumed to be binary and easily determinable by an analysis of biological factors. Despite anthropological and medical studies to the contrary, society presumes an unambiguous binary sex paradigm in which all individuals can be classified neatly as male or female.' Greenberg argues that throughout U.S. history the law has failed to recognize intersexuals, in spite of the fact that 1 to 4 percent of the world's population is intersexed. That is, they do not fit neatly into unambiguous sex categories; 'they have some biological indicators that are *traditionally* associated with males and some biological indicators that are *traditionally* associated with females. The manner in which the law defines the terms *male, female,* and *sex* will have a profound impact on these individuals' (emphases added)." (Lugones, 2007, 194)

15. "Epistemological analysis is an analysis of the contexts of experiences, more precisely the analysis of the theoretical content of experiences. We are concerned only with the theoretical content of the experience, that is, with the possible knowledge that is contained in the experience. (The analysis is not an actual division: the experience remains what it is: the analysis takes place in the course of a subsequent consideration of the already past and hence no longer alterable experience; hence it is only an 'abstractive', conceptual analysis.)" (Carnap, 1967, 308)

1.6   PREMISES AND CONCLUSIONS

To say that the foregoing problems constitute significant reasons for us to be interested in the study of logic is to have certain expectations about what will happen in the context of argument. Even if the activity of argumentation is itself under investigation, it seems clear that the process will require that we expect conclusions to be asserted and reasons to be given in support of those conclusions. These expected activities can provide direction in developing some preliminary resources with which to understand the process of argumentation.

Consider the following:

> Logic is to the philosopher what the telescope is to the astronomer: an instrument of vision. If we would become astronomers, we must learn the use of the telescope—not impatiently, chafing to get at more important work, but systematically, finding a certain intrinsic interest in the mysteries of the instrument itself. ... The same is true for logic; if we try to learn it in a cursory fashion, grudging the time and thought we would rather put on metaphysical problems, constantly wondering whether we have not acquired enough logic for our purposes, we shall never see the world in its clear light. We must work with a genuine interest in our restricted, abstract subject if we want it to lead us naturally to philosophical topics, as indeed it will—to problems of epistemology, metaphysics, and even ethics. (Langer, 1967, 41)

Two things should be clear in this passage from *An Introduction to Symbolic Logic*, by Susanne K. Langer. First, the passage presents Langer's conclusions about the importance of logic as a subject matter. At the same time, the manner of presenting her conclusions—her argument—is also aimed at leading the reader through a process of thinking that will let the reader share her conclusion. What is the conclusion? That we must study logic carefully. In offering the claim, Langer already supposes that her readers will call for reasons that support it. In this case, she constructs an analogy comparing the place of telescopes in astronomy with the place of logic in philosophy. Since astronomy requires the mastery of the technical skills of telescope use in order to properly study the stars, if the analogy holds, it follows that one must master the skills of logic in order to study the questions of philosophy: epistemology, metaphysics, and ethics. If we follow the steps presented by Langer in the passage, then we should at least be able to see how she justifies her understanding if not come to the same conclusion ourselves.

For purposes of talking about the activities that are basic to argumentation, we can begin by thinking of an *argument*, following Peirce, as literally any "process of thought that reasonably tend[s] to produce belief," which includes both arguments that respond to a situation at hand by producing guesses or hypotheses (Peirce calls this argument form *abduction*) and arguments that proceed "from definitely formulated premises" (6.456). Notice in the case of Langer's argument, the analogy between logic and telescopes is one that seems to emerge from no particular premises, but rather seems more like a creative development in thinking and talking about logic. Once the analogy is proposed, however, its characteristics become premises for making conclusions about how and why we should study logic.

The first step in understanding any argument is to determine the premises it includes and the conclusions it offers. When you consider examples of argument, in each case you can try to restate the premises and conclusions in as clear and concise a way as you can. Once you have done this, you can begin to think about two further issues. First, what situation would appear to serve as the starting point for the premises? If the premises are themselves conclusions of still other premises, what are these? If the premises, like Langer's analogy, do not seem to follow from other premises but from a creative act in a particular context, are they "successful" as an expression of the situation from which they emerge?

In the context of an argument, the premises and conclusions have the form of claims or *assertions* that say something about (serve as signs for) some definite state of affairs or definite sequence of actions. Such assertions may be "true" or "false" (without, at this point, worrying about the particular theory of truth adopted). In this case, 'Sue is an honors student' is a assertion while 'Shut the door!' is not (though it is easy to see that the command may also lead to assertions about a particular state of affairs or course of action).[5]

Further, an argument is the expression of an inference from premises taken to be true to a conclusion that must be true. This character of an argument is called *validity* and it underlies the expectation that an argument can be justified and can serve to bring others to share a particular conclusion. If the argument is valid and the speaker and her audience find a set of premises true, then they should share the conclusion as well. In case someone offers a series of assertions as an argument but where the premises lead to a conclusion in a way that is not valid (i.e., where the premises are true and the conclusion may be false), then we say that they have offered an *invalid argument.*

Valid arguments can also be understood as sound or unsound. A *sound argument* is one that has both a valid form and premises that are "true" in accord with some shared conception of truth. The claim in 2009 that Barack Obama is the current President of the United States is true for most people in that Obama occupies the White House, was elected by the Electoral College, and so on. Someone may also claim "If Barack Obama is the current President of the United States, then there was substantial election fraud in the last election." To conclude from these two claims that there was election fraud would be a sound argument if both premises are true. If, on the other hand, one or both of the premises are not true, the argument would remain valid, but not sound. For our purposes, an *unsound argument* will refer to arguments that are valid where one or more of the premises are false—again by whatever theory of truth is shared among those interested in the argument. It is possible, of course, to have an *invalid argument* with premises and a conclusion that are all taken to be true. For example, one might say "If Barack Obama is President, then he was elected by a majority of the American people" and, further, "Barack Obama was elected by a majority of the American people." One might then conclude rightly that Barack Obama is President. The argument (as we will see later) is invalid, but it is true that Obama is President and that

---

[5] Note that there is a difference between using an expression and mentioning it. An expression that is mentioned is surrounded by single quotes so that it is correct to say that 'Sue' is a three letter word while Sue is an honors student. It is also correct to say that 'Sue is an honors student' is an assertion (a claim about an expression) while Sue is an honors student (a claim about Sue). I will use this convention throughout: an expression in single quotes is mentioned rather than used.

he was elected by a majority of the American people in 2008. In this case, the argument is not unsound, it is invalid.

Consider the following argument from Aristotle's *On Interpretation* (Aristotle, 1941, 18b35–19a5). Here, Aristotle is concerned about the consequences of the view that for any statement, either that statement or its contradictory (that is, the statement negated) is true.

> [It] makes no difference whether people have or have not actually made ... contradictory statements. For it is manifest that the circumstances are not influenced by the fact of an affirmation or denial on the part of anyone. For events will not take place or fail to take place because it was stated that they would or would not take place, nor is this any more the case if the prediction dates back ten thousand years or any other space of time. Wherefore, if through all time the nature of things was so constituted that a prediction about an event was true, then through all time it was necessary that that should find fulfillment; and with regard to all events, circumstances have always been such that their occurrence is a matter of necessity. For that of which someone has said truly that it will be, cannot fail to take place; and of that which takes place, it was always true to say that it would be.

What is the conclusion of the argument Aristotle presents? In this example, the conclusion appears to be the statement "It makes no difference whether people have or have not actually made contradictory statements." If we were to restate this argument in order to make the premises and conclusion more explicit, we could write the following (although there are several ways of stating the argument).

Premise 1: Events will not take place or fail to take place because someone states that the events will or won't take place.

Premise 2: So circumstances are not influenced by the fact of an affirmation or denial on the part of anyone.

Conclusion: So it makes no difference whether people have or have not actually made contradictory statements.

The premises are related. The second premise is stated first by Aristotle and addresses the issue of making a statement that affirms or denies an event. If the premise is true, Aristotle will conclude that stating contradictory statements will make no difference. The second premise, however, depends upon a more general claim about circumstances (or events). So Aristotle offers what I have listed as the first premise as support for the second. I have here reordered them so that the first leads to the second and the second leads to the conclusion Aristotle proposes.

But what of Aristotle's discussion of prediction? Is this part of the argument? While the statement on prediction is made in support of what I have listed as the first premise, it is offered as a possible objection to the premise. It asks, "what if the prediction in question is made long before the event?" Aristotle's reply is that if a prediction was true in the past, it will be true at any time. But what support does this lend to the premise? Aristotle would probably answer that since a true statement is true always (and a false statement is always false), the fact that someone actually makes the statement at a given time and place has no bearing on the truth of the statement (or the occurrence of the event it regards). In this case, two premises are offered as a means of coming to share Aristotle's conclusion. Criticism of the argument (and indeed understanding the argument), turns on one's ability to determine the components of the argument.

1.7 EXERCISES

Consider each of the following arguments and restate them, in the simplest terms you can, giving the premises and conclusion of each. The goal, in each case, is to try to get at the argument the author is making by restating the premises and conclusion as clearly as possible. This is a common practice in philosophical investigation, but it is also a common practice whenever one is trying to be clear about another's efforts to present and support a conclusion. The following arguments are diverse in their points and approaches, but each is an effort to bring a reader to believe a particular conclusion.

1. "Learn, therefore, that the universe is not bounded in any direction. If it were, it would necessarily have a limit somewhere. But clearly a thing cannot have a limit unless there is something outside to limit it, so that the eye can follow it up to a certain point but not beyond. Since you must admit that there is nothing outside the universe, it can have no limit and is accordingly without end or measure. It makes no odds in which part of it you may take your stand: whatever spot anyone may occupy, the universe stretches away from him just the same in all directions without limit." (From Lucretius, 1966, 21, an Epicurean poet who lived in the first century BCE.)

2. "That a universal is not a substance ... can in the first place be evidently proved as follows: No universal is a substance that is single and numerically one. For if that were supposed, it would follow that Socrates is a universal, since there is no stronger reason for one singular substance to be a universal than for another; therefore no singular substance is a universal, but every substance is numerically one and singular. For everything is either one thing and not many, or it is many things. If it is one and not many, it is numerically one. If, however, a substance is many things, it is either many singular things or many universal things. On the first supposition it follows that a substance would be several singular substances; for the same reason, then, some substance would be several men; and thus, although a universal would be distinguished from one particular thing, it would yet not be distinguished from particular things. If however, a substance were several universal things, let us take one of these universal things and ask 'Is this one thing and not many, or is it many things?' If the first alternative is granted, then it follows that it is singular; if the second is granted, we have to ask again 'Is it many singular or many universal things?' And thus either this will go on *in infinitum*, or we must take the stand that no substance is universal in such a way that it is not singular. Hence, the only remaining alternative is that no substance is universal." (William of Ockham, 1964, 498–99, from a treatise on epistemological problems.)

3. "Brother, I rise to return you the thanks of this nation, and to return them back to our ancient friends—if any such we have—for their good wishes toward us in attempting to teach us your religion. Inform them we will look into this matter. We have well weighed your exertions, and find your success not to answer our expectations. But instead of producing that happy effect which you so long prom-ised us, its introduction so far rendered us uncomfortable and miserable. You have taken a number of our young men to your schools. You have educated them and taught them your religion. They have returned to their kindred and color, neither white men nor Indians. The arts they have learned are incompatible with the chase, and ill adapted to our customs. They have been taught that which is useless to us. They have been made to feel artificial wants, which never entered the minds of their brothers. They have imbibed, in your great towns, the seeds of vices which were unknown in the forest. They become discouraged and dissipated—despised by the Indians, neglected by the whites, and without value to either—less honest than the former, and *perhaps* more knavish than the latter.

"Brother, we were told that the failure of these first attempts was attributable to miscalculation, and we were invited to try again, by sending others of our young men to different schools, to be taught by different instructors. Brother, the result has been invariably the same. We believe it wrong for you to attempt further to promote your religion among us, or to introduce your arts, manners, habits and feelings. We believe that it is wrong for us to encourage you in so doing. We believe that the Great Spirit made the whites and the Indians, but for different purposes." (A speech given in 1805 at Buffalo Creek in New York by Sagoyewatha, a Seneca leader, 1866, 288–91. The speech was part of a conversation with a Christian mis-sionary who was meeting with the tribal council to request permission to evangelize in the Seneca community.)

4. "A man's power to connect his thought with its proper symbol, and so to utter it, depends on the simplicity of his character, that is, upon his love of truth, and his desire to communicate it without loss. The corruption of man is followed by the corruption of language. When simplicity of character and the sovereignty of ideas is broken up by the prevalence of secondary desires, the desire of riches, of plea-sure, of power, and of praise—and duplicity and falsehood take place of simplicity and truth, the power over nature as an interpreter of the will, is in a degree, lost; new imagery ceases to be created, and the old words are perverted to stand for things which are not; a paper currency is employed, when there is no bullion in the vaults. In due time, the fraud is manifest, and words lose all power to stimulate the understanding or the affections." (Ralph Waldo Emerson [in Harris et al., 2002, 49], from the essay *Nature*, published in 1836.)

5. "So the human brain has grown, by normal use and exercise, in the male; and been stunted, denied normal use and exercise in the female; and the more normal brain has been injured by the unavoidable contact with the less normal, and the race-brain—the common inheritance of us all—has been steadily robbed and weakened by the injury to the mother-half. ... The one bald fact that women have been kept shut up, confined to the house, in varying degrees of imprisonment, while men have been free to move, both locally and abroad, is enough to explain most of that 'subtle, baffling, wonderful' difference. If men had been arbitrarily divided into two classes; if half had been given the range of all trades, crafts, arts and sciences, all education, all experience, and achievement, and the other half shut up at home, never allowed out alone, and taken under escort to only a small part of life's attrac-

tions; confined exclusively to a few primitive industries; denied education as well as experience, we would find mysterious, subtle and baffling differences between the two classes." (Charlotte Perkins Gilman [Harris et al., 2002, 131], from *Our Brains and What Ails Them*, published in 1912.)

6. "So long as the mental and moral instruction of man is left solely in the hands of hired servants of the public—let them be teachers of religion, professors of colleges, authors of books, or editors of journals or periodical publications, dependent upon their literary labors for their daily bread, so long shall we hear but half the truth; and well if we hear so much. Our teachers, political, scientific, moral, or religious; our writers, grave or gay, are compelled to administer to our prejudices, and to perpetuate our ignorance. They dare not speak that which, by endangering their popularity, would endanger their fortunes. They have to discover not what is true, but what is palpable: not what will search into the hearts and minds of their hearers, but what will open their purse strings. They have to weigh every word before they pronounce it, lest they wound some cherished vanity, or aim at some favorite vice." (Frances Wright [Harris et al., 2002, 172–3], from a lecture given in 1834 to a working class audience in Philadelphia.)

7. "The reason the white man and Indian see two different orders of reality while gazing out over the same piece of land stems from a theory of knowledge about land which, in turn, rests on certain metaphysical presuppositions as to what land is and in what way it is. These different orders of perceived reality, in turn, give rise to two different experiences. The white man would say 'There is nothing here.' The Indian would say 'Everything is here.' It seems there is absolutely no agreement on what is 'here' even though both see the same phenomena. To the white man untilled land, an absence of towns or houses, is 'empty.' As nature abhors a vacuum, the white man abhors what he perceives as an empty void and hastens to 'fill it up' with fields, houses, barns, towns, and people. The vast emptiness of what was called the 'great American Desert' was strangely unsettling to the white pioneers. It was unbroken, rolling grassland stretching to the Rockies, and over this endless expanse blew a constant wind, at times softly but more often a shrieking wind, howling about the solitary cabins and farmhouses on the prairie. Stories are legion about settlers' wives going insane from a combination of space-drunkenness and howling winds. ... The Indian, on the other hand, perceives the land as a bountiful mother who provides everything he needs. The prairie is not an empty void between the Mississippi and the Rockies but is 'full' with every variety of living being, plants, animals, and people. Everything is there for the taking, it does not have to be tilled or worked or coaxed to provide for its children." (Robert Bunge, 1984, from his book, *An American Urphilosophie: An American Philosophy, BP (Before Pragmatism).*)

8. "The [first] doleful winter [experienced by the Pilgrims at Plymouth] broke up sooner than was usual. But our crippled planters were not more comforted with the early advance of the Spring, than they were surprised with the appearance of two Indians, who in broken English bade them, *welcome Englishmen!* ... This Indian [Squanto] ... by his acquaintance with the English language, fitted for a conversation with them, he very kindly informed them what was the present condition of the Indians; instructed them in the way of ordering their Corn; and acquainted them with many other things, which it was necessary for them to understand. ... This very strange disposition of things, was extremely advantageous to our distressed planters: and who sees not herein the special *providence* of the God

who disposeth all?" (Cotton Mather, 1853, Chapter 2, section 11, from *Magnalia Christi Americana: The Ecclesiastical History of New England.*)

9. "Whether peddled by white shamans, plastic medicine men and women, opportunistic academics, entrepreneurs, or enterprising New Agers, Indian spirituality—like Indian lands before it—is rapidly being reduced to the status of a commodity, seized and sold. Sacred ceremonies and ceremonial objects can be purchased at weekend medicine conferences or via mail order catalogs. How-to books with veritable recipes for conducting traditional rituals are written and dispensed by trade publishers. A succession of born-again medicine people have—with greater or lesser subtlety—set themselves and their services up for hire, ready to sell their spiritual knowledge and power to anyone willing and able to meet their price. And a literary cult of Indian identity appropriation known as white shamanism continues to be practiced. Instead of contributing to the many Native-run organizations devoted to enhancing the lives and prospects of Indian people, New Agers are regularly enticed into contributing to the continued expropriation and exploitation of Native Culture by purchasing an array of items marketed as means for enhancing their knowledge of Indian spirituality." (Laurie Ann Whitt, 1998, 140–1, from "Cultural Imperialism and the Marketing of Native America.")

10. "Truth [from a relativist position] is something like 'truth for me.' Call this *simple relativism.* Here's why simple relativism is so simple to refute. Suppose I am such a relativist and announce that there is no such thing as truth *per se*, there is only truth-for-me or truth-for-you. A fair question to ask would be whether the statement I just made is true or just true-for-me. If I say that relativism is simply true, then I have apparently contradicted myself. For if relativism is true (for everyone, as it were) then it is false—it is not true that all truth is relative. On the other hand, if I go the other way and say that relativism is only true relative to me, I am consistent but unable to convince anyone who doesn't already agree with me. You need only remark that relativism, then, is either contradictory or terminally unconvincing." (Michael P. Lynch, 2004, in *True to Life: Why Truth Matters.*)

11. "Feminist discourse on the Third World that assumes a homogeneous category—or group—called women necessarily operates through the setting up of originary power divisions. Power relations are structured in terms of a unilateral and undifferentiated source of power and cumulative reaction to power. Opposition is a generalized phenomenon created as a response to power—which, in turn, is possessed by certain groups of people. The major problem with such a definition of power is that it locks all revolutionary struggles into binary structures—possessing power versus being powerless. Women are powerless, unified groups. If the struggle for a just society is seen in terms of the move from powerlessness to power for women as a group, and this is the implication in feminist discourse that structures sexual difference in terms of the division between the sexes, then the new society would be structurally identical to the existing organization of power relations, constituting itself as a simple inversion of what exists. If relations of domination and exploitation are designed in terms of binary division—groups that dominate and groups that are dominated—then surely the implication is that the accession to power of women as a group is sufficient to dismantle the existing organization of relations. But women as a group are not in some sense essentially superior or infallible. The crux of the problem lies in that initial assumption of women as a homogenous group or category ('the oppressed'), a familiar assump-

tion in Western radical and liberal feminisms." (Chandra Talpade Mohanty, 2004, 38–9, from *Feminism Without Borders: Decolonizing Theory, Practicing Solidarity*.)

12. "Although the wonderful developments of human history teach that the grosser physical differences of color, hair and bone go but a short way toward explaining the different roles which groups of men have played in Human Progress, yet there are differences—subtle, delicate and elusive, though they may be—which have silently but definitely separated men into groups. While these subtle forces have generally followed the natural cleavage of common blood, descent and physical peculiarities, they have at other times swept across and ignored these. At all times, however, they have divided human beings into races, which, while they perhaps transcend scientific definition, nevertheless, are clearly defined to the eye of the Historian and Sociologist.

   "If this be true, then the history of the world is the history, not of individuals, but of groups, not of nations, but of races, and he who ignores or seeks to override the race idea in human history ignores and overrides the central thought of all history. What, then, is a race? It is a vast family of human beings, generally of common blood and language, always of common history, traditions and impulses, who are both voluntarily and involuntarily striving together for the accomplishment of certain more or less vividly conceived ideals of life." (W. E. B. Du Bois, 1897, from his essay "On the Conservation of Races.")

13. "The slightest reflection shows that the conceptual material employed in writing history is that of the period in which a history is written. There is no material available for leading principles and hypotheses save that of the historic present. As culture changes, the conceptions that are dominant in a culture change. Of necessity new standpoints for viewing, appraising and ordering data arise. History is then rewritten. Material that had formerly been passed by, offers itself as data because the new conceptions propose new problems for solution, requiring new factual material for statement and test." (John Dewey, 1938, 232, in *Logic: The Theory of Inquiry*.)

14. "ANARCHISM [is the] philosophy of a new social order based on liberty unrestricted by manmade law; the theory that all forms of government rest on violence, and are therefore wrong and harmful, as well as unnecessary. ... Anarchism is the only philosophy which brings to man the consciousness of himself; which maintains that God, the State, and society are non-existent, that their promises are null and void, since they can be fulfilled only through man's subordination. Anarchism is therefore the teacher of the unity of life; not merely in nature, but in man. There is no conflict between the individual and the social instincts, any more than there is between the heart and lungs: the one the receptacle of a precious life essence, the other the repository of the element that keeps the essence pure and strong. The individual is the heart of society, conserving the essence of social life; society is the lungs which are distributing the element to keep the life essence—that is, the individual—pure and strong." (Emma Goldman, 1911, from her pamphlet, *Anarchism: What It Really Stands For*.)

15. "The diverse crises suffered by Western culture have had their origin in the lack of coordination between the world of ideas and that of reality. When ideas fail to justify reality, the latter loses its human sense and it is necessary, then, to search for new ideas, new values, that will justify reality once again ... This should be the

task of the intellectual ... Otherwise, yesterday's living ideas turn into artifacts, into something dead or useless, without sense for the common man. The intellectual who continues toying with such ideas without trying to give them a new meaning is like a maniac, the possessor of a useless instrument." (Leopoldo Zea, quoted in Schutte, 1993, 92, from his book, *America como conciencia*.)

16. "In sharing a language, in whatever sense this is required for communication, we share a picture of the world that must, in its large features, be true. It follows that in making manifest the large features of our language, we make manifest the large features of reality. One way of pursuing metaphysics is therefore to study the general structure of our language. This is not, of course, the sole true method of metaphysics; there is no such. But it is one method, and it has been practiced by philosophers as widely separated in time or doctrine as Plato, Aristotle, Hume, Kant, Russell, Frege, Wittgenstein, Carnap, Quine, and Strawson. These philosophers have not, it goes without saying, agreed on what the large features of language are or on how they may best be studied and described: the metaphysical conclusions have in consequence been various." (Donald Davidson, 2001, 199, in his essay, "The Method of Truth in Metaphysics.")

17. "Perhaps our gifted author [John Locke] will not entirely disagree with my view. For after devoting the whole of his first book to rejecting innate illumination, understood in a certain sense, he nevertheless admits at the start of his second book and from there on, that ideas which do not originate in sensation come from reflection. But reflection is nothing but attention to what is within us, and the senses do not give us what we carry with us already. In view of this, can it be denied that there is a great deal that is innate in our minds since we are innate to ourselves, so to speak, and since we include Being, Unity, Substance, Duration, Change, Action, Perception, Pleasure, and hosts of other objects of our intellectual ideas?" (Gottfried Leibniz, *New Essays on Human Understanding*, 1765, paragraphs 44–53)

18. "Philosophic problems can only be defined operationally, as those questions in which philosophic minds have become interested. Or, what amounts to the same thing, they are those problems which have succeeded in generating a philosophic response, and have given birth to philosophers and philosophical questions and answers. They emerge, the record reveals, whenever the strife of ideas and experiences forces men [and women] back to the fundamental assumptions in any field; when it compels [these people] to analyze them, to clarify, criticize, and reconstruct them. If this be the case—and one cannot explore the historical dimensions of philosophizing without having the conviction forced on one that it is so—then the problems that initiate philosophizing have varied from age to age. They are to be understood ultimately only as expressions of the basic conflicts within a culture that drive men to thoroughgoing analysis and criticism." (John Herman Randall, Jr., 1963, 17–8, in *How Philosophy Uses Its Past*.)

19. "But any society is apt to be so viable that things are always shifting and are constantly being influenced by the larger community of the country and the world. The blame for unwelcome change can easily be ascribed to the influence or activity of the outsider. The common cry in the South [in the 1950's and 1960's] is that everything was moving along quite peacefully until the agitator from the outside interfered. The notion that segregated persons would ever of themselves on their own volition want conditions to change is impossible to entertain. And the reason is not far to seek, for if it were ever admitted, then the whole structure of the

relationship would be brought into question. This would create an intolerable situation. From the point of view of power, the stability of the order rests upon its total acceptance. To this end, the ancient role of the scapegoat is called into play. The mind of the South is temporarily incapable of dealing with dissatisfied and hostile restless Negroes. To admit this reality is to destroy the myth; to destroy the myth is to have all landmarks shift, making it impossible to hold one's course and keep one's bearings. Hence every effort has to be made to keep out the outsider and at the same time to open the way for the outsider to enter that he may be available to bear the brunt of the cause for unrest." (Howard Thurman, 1965, 9, in *The Luminous Darkness*.)

20. "I never react to you but to you-plus-me; or to be more accurate, it is I-plus-you reacting to you-plus-me. 'I' can never influence 'you' because you have already influenced me; that is, in the very process of meeting, by the very process of meeting, we both become something different. It begins even before we meet, in the anticipation of meeting. We see this clearly in conferences. Does anyone wish to find the point where the change begins? He never will. Every movement we make is made up of a thousand reflex arcs and the organization of those arcs began before our birth. On the physiological, psychological and social levels the law holds good: response is always to a relating. Accurately speaking the matter cannot be expressed even by the phrase used above, I-plus-you meeting you-plus-me. It is I plus the-interweaving-between-you-and-me, etc., etc." (Mary Parker Follett, 1924, 62–3, in her book *Creative Experience*.)

WHAT IS LOGIC?

2.1   THE STUDY OF LOGIC

The story of "Modern" logic can start in a variety of places. Some hold that the best place to begin is in the work of Immanuel Kant, whose *Critique of Pure Reason* challenged philosophers to reconsider the place of logic in human understanding and whose theory of logic gave rise to competing approaches. In his *Introduction to Logic* (notes he used in teaching his regular logic class) Kant writes: "This science of the necessary laws of the understanding and reason in general, or—which is the same—of the mere form of thinking, we call *logic*" (Kant, 1974, 15–6).

For Kant, logic was the study of the process of thought in its ideal sense, not as it is practiced. "Some logicians," he says, "presuppose psychological principles in logic. But to bring such principles into logic is as absurd as taking morality from life" (1974, 16). The comparison with morality is important. For Kant, logic is not about how people happen to think, but rather what is necessary in order to think well. From this angle, logicians are therefore concerned in equal measure with finding necessary truths and providing a standard according to which the actual practice of people can be evaluated.

Kant's notion of logic as "*the science of thought*" historically led to a division in logic: one branch treating logic as the study of the operations of mind in the process of knowing and the other having to do with the conditions of knowledge independent of thought, that is, logic as concerned with truth. The former approach considered the process of thought as a matter of the ways in which ideas relate to one another. To come to the conclusion that an apple is on the table, one would first need to have certain ideas about apples and tables. Logic is concerned with the ways in which some of these ideas are related to certain other ideas so that, given ideas about tables and apples, the idea of an apple on the table follows. Further, according to this view, since humans only have access to ideas about the world (i.e., we never grasp apples and tables themselves but only ideas of them), logic is finally only about ideas—only about thought. Such a view of thinking could reveal (according to its advocates) general patterns of relations among ideas such that they could guarantee certain beliefs, but it also necessarily became a process of theorizing about how the mind, rather than the world, worked.

By the end of the 19th century, many argued that the concept that logic was strictly about the "laws of thought" was really a form of psychology. Since Kant rejected the consideration of logic in relation to particular experience, it was a form of psychology that had little to do with knowledge and truth as it was found in the rapidly developing fields of the physical sciences. To understand scientific knowledge, many logicians decided they should stop treating beliefs as a matter of mental relations and begin to understand thought in terms of the relations that made them true. Rather than being concerned with the laws of thought, logic should consider the laws of truth as they emerge in mathematics, geometry, and physics. Rather than analyzing the interaction of ideas in consciousness, these theorists wanted to understand the relation of claims independent of consciousness so that, given certain claims about the world, certain other claims correctly follow. In geometry or arithmetic, when given the truth of certain axioms, certain theorems follow by necessity. This was not a matter of the rules of thought, but rather a matter of rules independent of thought. Science, these logicians argued, is a process of finding truths in the world, not a matter of studying the relations among the ideas of science.

This approach from the perspective of "laws of truth" led to two intertwined concerns: What makes claims about the world true? And what are the rules that allow us to go from one set of true claims to another set using thought alone? The first question became the subject matter of epistemology, which sought to explain how we come to grasp truths about the world. Logic became primarily concerned with the second interest, that is, an interest in the rules of inference that preserve truth even as we infer new claims about the world. This conception of logic is well described by W. V. O. Quine in the 20th century as *"the science of necessary inference"* (1965, 1). What is important on this model is not the operation of the mind, but the "truth-preserving" relations among statements about the world.

The result of this interest in truth preservation is a theory of inference that holds regardless of the particular content of statements and regardless of the particular operations of the mind. As Morris Cohen put it in his *Preface to Logic*, "the distinctive subject matter of logic … is what is called formal truth" (Cohen, 1944, 3). There is, he argues, a difference between the truth of a statement such as "There is an apple on the table" and "If there is an apple on the table, then there is a table under the apple." The first is true if there is an apple on the table. This is a contingent state of affairs so that the statement might be true in some cases and false in others. The second, however, is true regardless of the situation. It is a formal truth, Cohen argues, because it is the form of the statement that guarantees its truth and not any particular situation. On one hand, this is exactly the sort of thing Kant was interested in as the subject matter of logic—necessary truth. On the other hand, as logic developed, it was not about thought but rather about necessary relations independent of thought.

A third approach to logic accepts that both the operations of thought and the formal "truth-preserving" structure of statements are important, but reframes these ideas within what is often called the context of *inquiry*. C. S. Peirce's remark that the process of drawing inferences is "a long and difficult art" is instructive. As with any art, the artist uses a set of tools and other resources that work in more or less particular ways. Two musicians can produce radically different music because they have different backgrounds, different physical capabilities and different attitudes about music. At the same time, the range of tones available for a musical piece is more or less the same regardless of the musician. To perform music involves both the mind of the artist and the "formal" relations of music. Just as the art of musical performance involves things independent of a particular artist and things which are distinctive to the talent and mind of a particular artist, so the process of reasoning can be seen as involving both the structured relations of truth and the operations of mind, or rather, the activities and dispositions of individual inquirers. For purposes of this discussion, the process of reasoning and its concern with the relations among ideas and conditions of truth will be called *inquiry*.

From this third perspective, logic is the study of the activity of inquiry as it can be generalized over a wide range of cases. The principles that result from this general study amount to both descriptions of the practice of logic and guidelines for the further conduct of inquiry. These are developed in the context of actual inquiries and are subject to revision as new inquiries occur. The resulting logic—now called the theory of inquiry—is a "logic" described by John Dewey as having six implications:

1. "*Logic is a progressive discipline*," that is, as a "science" logic develops over time as theories change and resources for investigation evolve. In effect, logic is an evolving set of theories.

2. "*The subject-matter of logic is determined operationally*." What one studies in logic is what one finds in the context of actual inquiry. As a result, "formal" elements of the process (including logical operators such as 'and', 'or', 'if/then' and so on, as well as systems of measurement and computation) along with search procedures, experimental practices, even those structures that enable and constrain inquiries systematically are all potentially part of logical subject matter.

3. "*Logical forms are postulational*." In this case, the structures or principles which "govern" the operations of formal elements of inquiry are "postulates" in the sense that they are "a generalization of the nature of

the means that must be employed" if "true" statements are to be made and related. But postulates according to Dewey are not only descriptive of the process of inquiry, they also are normative. "To engage in an inquiry," Dewey observes, "is like entering into a contract. It commits the inquirer to observance of certain conditions." These conditions might involve particular practices in conducting experiments (such as procedures for record-keeping and preparing microscope slides) or practices regarding interactions with others (such as asking and answering questions, giving reasons, and submitting results to others for confirmation).

4. *"Logic is a naturalistic theory."* Here "naturalistic" has two meanings: "on one side, that there is no breach of continuity between biological operations and physical operations. 'Continuity' on the other side means that rational operations grow out of organic activities, without being identical with that from which they emerge."

5. *"Logic is a social discipline."* In other words, logic is concerned (as already suggested) with the social and cultural context in which inquiry is carried out and in which the "principles" of logic are established, questioned and revised. "Inquiry is a mode of activity that is socially conditioned and that has cultural consequences."

6. *"Logic is autonomous."* Dewey explains this way: "Logic as inquiry into inquiry is, if you please, a circular process; it does not depend upon anything extraneous to inquiry. The force of this proposition may be most readily understood by noting what it precludes. ... It precludes resting logic upon metaphysical and epistemological assumptions and presuppositions. The latter are to be determined, if at all, by means of what is disclosed as the outcome of inquiry. ... Knowledge is to be defined in terms of inquiry, not vice-versa... " This last implication raises interesting questions about the place of knowledge and ontology relative to logic. It implies not that there is no such thing as metaphysics or epistemology, but rather that there can be no ontology or knowledge without the ongoing processes that serve as the subject matter for logic. Put another way, metaphysics and epistemology are an outcome of inquiry and so the nature of "being" and knowledge cannot be separated from the processes of inquiry. Logic is "autonomous" in that it does not rest on prior or antecedent principles, but it is not independent of other studies either, since the products of past inquiries into being and knowledge frame any inquiry into inquiry.[1]

The resulting conception of logic is one in which a wide range of questions play a role. Many students begin the study of logic thinking that it will be primarily concerned with formal systems. Like systems of mathematics, students expect to learn a set of rules and then, over the course of a term, study the ways the rules can be used to prove various theorems of the system.

[1] All quotations above are from Dewey, 1938, 21–9.

With the skill of carefully applying rules, students expect to benefit indirectly by learning a formal system even if it has little to do with daily concerns. Those with a broader interest in philosophy might see the study as a kind of preparation for other work in philosophy, even as they expect little that is "philosophical" from the study of logic itself.

Other students come to logic expecting something practical. Logic, they say, is not about memorizing and applying abstract rules but learning to argue effectively and becoming an incisive critic who can identify and destroy the bad arguments of others. Such students often do not wish to be troubled by the details of formal systems nor are they particularly interested in philosophical matters. I caution both students interested in "pure" formalism and those interested in practical applications. The systems of order that emerge in the study of inquiry can be formalized to a certain extent and these "formalisms" can tell us much about how we think and live. To study logic from the angle of vision that I propose here is to take formal relations seriously, to learn something about proofs, and also to learn something about the philosophical import of relations such as implication and negation. At the same time, formal systems get their meaning from the arguments in which they are used. Political arguments, scientific investigations, court trials, literary criticism, and even philosophical analysis are all instances of inquiry. Just as a musician benefits from knowing her instrument and harmonic structures, so an inquirer benefits from both *knowing about* inquiry and *practicing* it.

2.2   THE CONCEPTS OF TRUTH AND INFERENCE

While it is perhaps clear how this third conception of logic incorporates the idea of logic as the science of thought in its attention to the process of inquiry, it may remain unclear how it is also a science of "necessary inference" or of "truth preserving" rules. To see the place of truth, it is helpful to begin with traditional notions of truth. These traditional theories can be divided into three kinds: *correspondence* theories, *coherence* theories, and *deflationary* theories.

The first of these theories, *correspondence*, is perhaps the most obvious in that theories of this sort begin with the common sense notion that a statement is true if and only if whatever it describes is as it says. If we say that it is raining outside, our statement is true if it is, in fact, raining outside. Statements are said to be true when they "correspond" to the thing or state of affairs that is their referent. Such a notion of truth often runs into opposition, however, when it becomes apparent that diverse statements about the world require more than just correspondence to be true.

One version of the correspondence theory, important in the development of logical theory in the 20th century, is called the *identity theory* of truth.[2]

---

[2] Bertrand Russell (1938) in his early work and Gottlob Frege (1918) each held versions of the identity theory.

This approach is intended to close the gap between what is asserted and what makes it true. In this case, theorists observed that an assertion is true when it corresponds to a "fact." If the claim that Shirley Chisholm was the first black woman to run for U.S. president is true, then it is because it is a fact that Shirley Chisholm was the first black woman to run for president. The theory holds that truth is a matter of identity because what makes the assertion true is that it is itself a fact, that is, the "truth bearer" (the assertion) is identical with the "truth maker" (the fact). The problem faced by identity theories emerges when one tries to explain the nature of the facts. On one hand, if the facts are just whatever assertions one makes, then it is not clear how the facts connect with the world that the assertions are to describe. On the other hand, if the facts are independent of their being asserted, then it appears that they must exist somewhere as a kind of unspoken assertion outside of the language practices of humans.

Whether the "truth makers" are taken to be facts or states of affairs, claims nevertheless seem to require more than simple correspondence. For example, the claim that the world was created in six days requires, among other things, some notion of days and creation. The old question from the play *Inherit the Wind* gets at the point: since days are defined in terms of the movement of the sun, what constituted a "day" before God said "Let there be light"? Further, it is not enough that a statement about a thing be true if it is not also compatible (or consistent) with the other true statements about the world. From this perspective, a statement is true if and only if it coheres with the system of which it is a part.

The *coherence theory* is easily understood if you imagine a series of claims about the state of the local weather at the moment you read this. Perhaps, it is cold. It is raining. The sky is overcast. But now suppose someone calls to tell me that it is sunny and dry. I don't even need to look out my window to tell him that he is mistaken. If it is true that it is raining and overcast, then these new claims to the contrary do not cohere. From this perspective, truth is less about the connection between statements and the world and more about their consistency in a system of statements.

The final group of truth theories, labeled *deflationary* theories, amount to the conclusion that truth is an unimportant property of statements and can safely be set aside. To see that this view is, in a sense, unproblematic, consider the difference between the statements "It is raining" and "It is true that it is raining". To most, these statements say the same thing. The addition of 'it is true' to the second version adds nothing to the meaning of the original. From this perspective, truth, as a special property of assertions, can be set aside and replaced with characteristics related to, for example, what justifies the claim.

Another version of this approach begins with the observation that assertions must be verified and that these verifications can be understood as the conditions under which the assertion is true. Alfred Tarski, in an effort to understand the truth conditions of formal languages, proposed that the truth conditions of a language could be expressed in a series of assertions about the language. The assertion of these truth conditions would take the following form: an assertion, S, in a given language is true if and only if (iff) certain conditions, C, hold. As a result, the truth conditions for an assertion would have the form:

S is true iff C.

In order to specify a form that will express the truth conditions of any assertion in the language, S will be replaced by a term that names each of the assertions of the language. One way to name an assertion, for example, the assertion that the sky is overcast, is by surrounding it with quotation marks. The quotation, "The sky is overcast", stands for the assertion that the sky is overcast. The assertion itself stands for (or points to) a state of affairs in which the sky is overcast. This can be formalized by saying that "p" names the assertion that p. Since p has the function of representing the thing or state of affairs described by "p", the resulting truth conditions have the general form:

"p" is true iff p.

Given this approach, the truth condition for the assertion, "Rhododendrons are green" would be

"Rhododendrons are green" is true iff rhododendrons are green.

The resulting list of truth conditions for the assertions of a language is called a *disquotational* theory of truth (because the truth condition for each assertion in the language is simply the assertion without quotation marks).

Hartry Field argues in favor of a deflationary disquotational theory of truth. He observes "[w]hat [the theory] is deflationist about is truth conditions" (1994, 253). What this approach avoids is what Field calls inflationism: the need to posit truth-making facts. "[A]n inflationist," Field concludes, "is apparently faced with the three options of finding … facts, or accepting it as simply a brute fact that our word stands for genuine identity, or accepting a surprising level of referential indeterminacy in our basic logical vocabulary. This is a choice that not everyone would be happy to have to make" (1994, 258). Yet, despite the problems deflationist theories avoid, they also set aside the aspects of truth explained by correspondence and coherence theories.

The notion of truth that operates in the approach taken in the theory of inquiry is that an assertion is true if it is the outcome of an inquiry. In this case, truth is not a fixed characteristic of a given statement since statements are only as "true" as the inquiries that produced them are "successful." At the same time, this does not imply that truth is "relative" in the sense that anything can be true if we want it to be or that any statement I find to be true in inquiry is simply "true for me" and not necessarily for you. If we take seriously what we will later call the "situated" character of inquiry, then results are highly constrained by the situations in which the inquiries are conducted. The theory of inquiry also recognizes a second notion of truth as a characteristic shared by claims *used* in inquiries. The results of past inquiries, even though they are generally open to further review are, at times, tools of inquiry and as such are taken as "truths" (i.e., postulates in further inquiry). These past results along with hypothetical claims not yet grounded in successful inquiry are nevertheless crucial to the conduct of inquiries and so are rightly understood or "taken" as true. We will return to this issue later.

From the perspective of most 20th century logic, the central concern is not what makes claims true in the first place, but rather what relations among claims govern the process of inference to "preserve" truth. Here, *"inference"* is defined as the process in which, taking certain claims as true, certain other claims follow. Peirce describes this process as a matter of habit. From this angle, the experience or grasp of some premise, *p*, by habit leads to a conclusion, *c*. "A habit is not an affection of consciousness," Peirce argues, "it is a general law of action, such that on a certain general kind of occasion a [person] will be more or less apt to act in a certain general way" (2.148). When one encounters certain claims, the habit of inference provides a kind of structured activity that leads to other claims that are said to follow from the first. While one may propose a great variety of connections as possible rules connecting premises and conclusion, Peirce argues that the ones that we tend to keep and foster, that is, the good habits, are ones with a definable characteristics. "The habit," he says, "is logically good provided it would never … lead from a true premise to a false conclusion; otherwise it is logically bad" (3.163).

Notice that Peirce proposes a standard that allows any connection to be tried as an inferential connection. Those that lead to true conclusions are ones that may be good and so ought to be preserved. Those that lead to false conclusions are bad and so ought to be abandoned or modified. The resulting view recognizes both a general activity of inference and an on-going process of adjustment that allows for experimentation and change. Peirce further argues that beliefs "are all adapted to an end, that of carrying belief, in the long run, toward certain predestinate conclusions which are the same for all men. … Hence, if a given habit, considered as determining an inference, is of such a sort as to tend toward the final result, it is correct; otherwise not. Thus, inferences become divisible into the valid and the invalid; and thus logic takes its reason of existence" (3.161).

Consider the claim:

1) If Jackie is late for class, she will miss hearing the assignment for the next class.

If we take this claim to be true (by whatever theory of truth we accept), and if we find (again by whatever theory we "find" such things) that

2) Jackie is late for class.

We infer that

3) Jackie will miss hearing the assignment for the next class.

Notice that the relation that underlies the inference is not particular to Jackie and her class. To see this, we can first symbolize the relations that are operative. Let 'J' symbolize the claim "Jackie is late for class" and 'M' symbolize

"Jackie will miss hearing the assignment." The first claim, (1), relates the two claims, J and M, to each other through the use of 'if' and 'then': If J then M. We can symbolize the if/then relation between the claims using '$\rightarrow$' so that 'J $\rightarrow$ M' can represent the initial complex claim. This complex claim is generally called a *conditional assertion* because it asserts that J and M are related in light of certain conditions. The first claim, J, is called the *antecedent* of the conditional (since it comes before the second claim) and M is called the *consequent* because it follows as a consequence of the first. The two claims we take to be true above (assertions (1) and (2)), along with the conclusion that must be true, (assertion (3)), can be represented symbolically as

(1′) J $\rightarrow$ M

(2′) J

(3′) $\therefore$ M[3]

Notice that this same form can be used to symbolize any number of claims. For example, if the assertion "If Henry goes to the movie tonight, he will not attend the study session" is taken as true and "Henry goes to the movie tonight" is also true, then we can infer that he will not attend the study session. The symbolization of this example and the one preceding it suggests a common form. This common form represents, in effect, the structure of an *inferential habit*. This can be symbolized by the following:

(i) p $\rightarrow$ q

(ii) p

(iii) $\therefore$ q

This form is usually called *modus ponens*.[4] While assertions such as "Henry goes to the movie tonight" are symbolized by an appropriate *uppercase letter*, 'H', for example, the form of the assertions given here uses *lowercase letters* to indicate that *any* assertion can stand in their place, not just the ones about Henry.

Finally, it is important to note that the symbolized form just presented has a further crucial property: whenever the *premises* (i and ii) are true, then the *conclusion* (iii) *must also be true*. This characteristic is called *validity*. Regardless of what specific claims p and q are, if they are related in the way specified by the conditional, and p is true, it must be the case that q is also true. Much of modern logic has been narrowly concerned with the truth preserving characteristics of valid argument forms. Philosophers concerned with

---

[3] The symbol '$\therefore$' or 'therefore' indicates that what follows is a conclusion that must follow from the claims that precede it.

[4] "*Modus ponens*" is a Latin expression meaning "mode of positing or affirming" in contrast to another form called *modus tollens* or "mode of taking away or negating." The form of *modus tollens* is as follows where "¬" means "not":

    (i) p $\rightarrow$ q

    (ii) ¬q

    (iii) $\therefore$ ¬p

logic in the American philosophical tradition, including Peirce and Dewey, took validity to be part of the wider process of inquiry.

2.3   THE PROCESS OF INQUIRY

In John Dewey's book *How We Think*, he offers a classic example of inquiry:

> Suppose you are walking where there is no regular path. As long as everything goes smoothly, you do not have to think about your walking; your already formed habit takes care of it. Suddenly you find a ditch in your way. You think you will jump it ...; but to make sure, you survey it with your eyes ..., and you find that it is pretty wide and that the bank on the other side is slippery. You then wonder if the ditch may not be narrower somewhere else ..., and you look up and down the stream ... to see how matters stand. ... You do not find any good place and so are thrown back upon forming a new plan. As you are casting about, you discover a log. ... You ask yourself whether you could not haul that to the ditch and get it across the ditch to use as a bridge. ... You judge that idea is worth trying, and so you get the log and manage to put it in place and walk across. (Dewey, 1933, 198)

In the example, Dewey asks us to imagine walking, perhaps across a high mountain meadow. So long as the walk is smooth and uninterrupted—and presuming you are just enjoying the stroll and the day—you are not involved in inquiry. When you encounter an obstruction, in this case a ditch or a steep streambed, you are suddenly brought up short.

Although Dewey does not say it, the moment you are stopped probably has a distinctive quality or "feeling" about it—surprise perhaps, or frustration, or simply a sense of being disrupted. As he puts it elsewhere, the "quality" of this sort of situation is one of disruption or frustration, where the quality is not properly just a matter of your feeling, but is actually a kind of *pervasive quality* that takes in the entire scene.[5] Peirce calls the "feeling" connected with moments of this sort "doubt" and Dewey calls it more generally "indeterminacy."[6] From this perspective, the quality is not simply a feeling, but a unique quality that marks the entire situation. As Dewey explains,

> Thus it is of the very nature of the indeterminate situation which evokes inquiry to be questionable; or, in terms of actuality instead of potentiality, to be uncertain, unsettled, disturbed. The peculiar quality of what pervades the given materials, constituting them a situation, is not just uncertainty at large; it is a unique doubtfulness which makes that situation to be just and only the situation it is. It is this unique quality that not only evokes the particular inquiry engaged in but that exercises control over its special procedures. ... A variety of names serves to characterize indeterminate situations. They are disturbed, troubled, ambiguous, confused, full of conflicting tendencies, obscure, etc.

---

[5] See Dewey, 1938, 74–5.
[6] See Dewey, 1938, 72–3.

For Dewey, the pervasive quality of indeterminacy characterizes the situation and marks the beginning of inquiry. In the case of the traveler, for instance, the indeterminacy is not just "in" the traveler, but rather is a matter of the situation that includes the traveler, her walk, the meadow, the ditch and so on.

> It is the situation that has these traits. We are doubtful because the situation is inherently doubtful. Personal states of doubt that are not evoked by and are not relative to some existential situation are pathological; when they are extreme they constitute the mania of doubting. Consequently, situations that are disturbed and troubled, confused or obscure, cannot be straightened out, cleared up and put in order, by manipulation of our personal states of mind. The attempt to settle them by such manipulations involves what psychiatrists call 'withdrawal from reality.' Such an attempt is pathological as far as it goes, and when it goes far it is the source of some form of actual insanity. The habit of disposing of the doubtful as if it belonged only to us rather than to the existential situation in which we are caught and implicated is an inheritance from subjectivistic psychology. ... Restoration of integration can be effected ... only by operations which actually modify existing conditions, not by merely 'mental' processes. (Dewey, 1938, 109–10)

In the face of an interruption of the sort faced when you encounter an obstacle, Dewey argues that you begin to try a variety of activities aimed more or less at eliminating or overcoming the obstacle in order to restore the smooth flow of the walk you began on the far side of the meadow. Put another way, when the pervasive quality of the situation becomes indeterminate, humans often respond by initiating a process of inquiry that aims at transforming the situation into one that is determinate, that is, it is organized enough to allow activity (the previous one or something new) to resume.

Dewey says that the function of the process needed by the traveler, what calls he "reflective thinking," is "*to transform a situation in which there is experienced obscurity, doubt, conflict, disturbance of some sort, into a situation that is clear, coherent, settled, harmonious*" (Dewey, 1933, 195). He defines the process of *inquiry* in general as "*the controlled or directed transformation of an indeterminate situation into one that is so determinate in its constituent distinctions and relations as to convert the elements of the original situation into a unified whole*" (Dewey, 1938, 108). The first definition captures the experienced quality of the process for the inquirer. It focuses on the experienced quality of a doubtful situation that pushes us to reflect and, if all goes well, to resolve the situation into a new harmony. The second definition stresses the character of inquiry as a process of situations as a whole. Despite our best efforts, inquiry is never a product of a thinker in isolation. Interests formed in our communities, the constraints of our environments, and the availability of resources are present from the beginning. Even in the case of the interrupted walk through a meadow, the traveler's interests and ways of thinking about problems inherited from her culture, are part of the situation and may contribute to (or block) the effort to transform it. The situation as a whole gives rise to an inquiry and it is the situation as a whole marked by its pervasive quality of determinacy that marks an inquiry's end.

In Dewey's analysis of the traveler's response to the situation, he observes that five distinct processes emerge. The first, "intellectualization" or (1) *problematization*," is the process of determining what kind of disruption or obstacle one faces. Is this a problem of crossing a ditch, taking a wrong turn, being lost in the mountains, or some other problem among an endless number of possibilities? Clearly the traveler's response will turn in a crucial way on how the situation is understood as a particular problem. In this case, the traveler decides that she faces the problem of getting across the ditch.

The next stage, (2) *suggestion*, is the process of casting about for a way to "restart" the walking. "[T]he tendency to continue acting," Dewey says, "nevertheless persists. It is diverted and takes the form of an idea or suggestion" (1933, 44). In this case, the traveler considers jumping the ditch and then using a log for a bridge.

The process of suggestion gives over naturally to the process of (3) *forming a hypothesis* that, in the next step of the process, is tested, not overtly by trying to jump, but by imagining the possibility. "I could jump here," she proposes, but then imagines trying the jump, and (with visions of crashing into the ditch) proposes another hypothesis: "perhaps I could cross somewhere else." When she realizes that there is no narrow place, she casts about again and proposes to use a log as a bridge. She imagines the possibility and concludes that it might work. In the next step of the process, she drags the log to the ditch to test its length and finally walks across. In effect, she (4) *tests the hypothesis first by "reasoning"* or in thought (or, as Dewey calls it elsewhere, "dramatic rehearsal") and then (5) *tests it by experience*, walking across the log to the other side.

In the process described, the traveler uses ideas and overt activities to address the indeterminate situation. The ideas proposed in the process, like the premises of inference discussed in the previous section, lead to conclusions that, as actions, may resolve the situation. In the process of inquiry, the premises include *hypotheses* generated by problematizing, suggesting and hypothesizing as well as *postulates*, that is, ideas and other instruments that are products of past inquiries. In the case of the traveler, the hypothesis that proposes jumping the ditch is tested by reason and rejected thanks in part to postulates about one's ability to jump long distances and the effects of a long fall. The hypothesis that the ditch can be crossed using a log is tried, in part, because of postulates about bridges and the strength of logs to hold the weight. Combined with the overt work of testing the log hypothesis, the situation is transformed and the traveler continues on her way.

As she steps off the log on the far side, the situation is transformed. Her walk is restarted and she feels a qualitative change in the situation as a whole. In the same way that one gets a satisfied feeling when one meets with a success, so the traveler's situation is transformed from one of disruption to one opened again to new possibilities, to the meadow, and to the plans that had been put on hold. The outcome has two aspects related to the two notions of truth discussed above. One is to restore the movement of the walk across the meadow. The hypothesis that the traveler could cross the ditch using a log as a bridge is confirmed—the obstacle is overcome and the journey is restored. The proposal that the ditch can be crossed using a log is true as the outcome of the inquiry. The second outcome, the true assertion, is a kind of byproduct. The traveler moves on but now with new resources that can help guide future walks. Not only does she know about the ditch across the meadow, she also knows something about crossing a ditch that could be used again in the future when a similar problem comes up.

In Dewey's description of the process of inquiry, once an inquirer finds herself in an indeterminate or problematic situation, the first response is to determine what sort of problem she faces and what, given that problem, she might do to transform the situation. Dewey describes this stage in vague terms: he writes that the process of suggestion is one "in which the mind leaps forward to a possible solution" (Dewey, 1933, 200). What constitutes this "leap" is not clear. Where does it begin? What determines its direction? What constitutes the connection? What such leaps amount to, Dewey says, is a kind of inference that begins in the present moment constrained by present circumstances and directs us toward a subsequent situation. Dewey concludes: "Every inference, just because it goes beyond ascertained and known facts, which are given either by observation or by recollection of prior knowledge, involves a jump from the known into the unknown. It involves a leap beyond what is given and already established" (1933, 191).

In his consideration of the process of inquiry, C. S. Peirce identifies the preliminary stages of inquiry (including problematizing, suggesting and hypothesizing) as different functions of the operation he calls "*abduction*." Peirce defines the operation this way: "Abduction is the process of forming an explanatory hypothesis" (5.171). The process is a crucial one because it "is the only logical operation which introduces any new idea" (5.171). Sometimes this process is better known as *guesswork* (7.219). Given a problematic situation, the inquirer makes a guess about how the situation might be transformed. Sometimes the guess pays off. Other times, the guess goes nowhere or even makes the situation worse in some way.

One might argue that guesswork is not part of inquiry, but, at best, is a kind of mindless trial and error process that brings possibilities into consideration at random. On the contrary, Peirce argues that abduction is rather the first form of reasoning because, unlike random trial and error, the process of abduction is already constrained by the situation at hand. When one makes a guess, it is rarely from among an array of possibilities so large that it includes anything at all. If Sarah asks Juan where her class is meeting and Juan does not exactly know the answer, he might guess. The guess he makes will probably not be that the class is being held in a romance novel or on the surface of the moon. Rather, he will guess that it will be held in a classroom, perhaps the one he just left or the one he is going to. Of course, Juan may be wrong about the location of the class. Still, it is the initial guess (arising from the situation at hand) that allows Sarah to continue her quest for her classroom. If

Juan had not answered (or had given an answer that had nothing to do with the situation), Sarah would have had no new ideas to try and her inquiry may grind to a halt unless, for example, she takes a guess or encounters someone else who will do so.

Put another way, abduction is a form of inquiry, but it is both necessary and the least reliable form. It relies (as we will see later) on only one premise—the situation at hand, proposes hypotheses, and, in light of the situation and the guess, justifies trying something that might transform the whole. The proposal of a particular premise, however, means that the process is not simply random selection. Instead, what is present in the situation can be said to actively suggest a claim. While whatever is at hand may not make definitive suggestions, it does make relevant ones. Inquirers who are sensitive to the situation will find that ideas "pop up" spontaneously. Just so, the best scientists attempt to be open to whatever a set of experimental results might suggest, even as the results are constrained by the circumstances of the laboratory and the resources available to the scientist. Peirce argues that this responsiveness to the situation amounts to learning from *il lume naturale*, "the light of nature" (6.477). What mathematicians find as a moment of insight in which difficult problems are solved, what poets find as a moment in which the right words emerge, and what religious people take as revelations of God can all be understood as inquiries in the simplest form—as abductions.

A NEGLECTED ARGUMENT

Peirce develops the notion of abduction in a instructive way in his paper, "A Neglected Argument for the Existence of God." While many philosophers have proposed proofs (e.g., ontological and cosmological arguments), Peirce argues that the process of abduction can likewise lead to a case for the existence of God. At its core, abduction is a process in which an inquirer is open to the possibilities suggested by the situation at hand. To the extent that one is already convinced of a particular idea or that a particular answer will do, her expectation will affect what, if anything, might "leap up" or be suggested as a response to an indeterminate situation. Instead, Peirce argues that "there is a certain agreeable occupation of mind which, from its having no distinctive name, I infer is not as commonly practiced as it deserves to be; for indulged in moderately … it is refreshing enough more than to repay the expenditure" (6.485). Peirce calls this occupation "Pure Play" which, he continues, "involves no purpose save that of casting aside all serious purpose, … has no rules, except this very law of liberty. It bloweth where it listeth" (6.458).

In its most open form, the process of abduction is one of "pure play" or "Musement"—a process of reflection that is highly sensitive to what one's situation will suggest. By setting aside as many constraints as possible and letting inquiry become free play, Peirce suggests, the process follows a "natural" course. "Let the Muser, after well appreciating, in its breadth and depth, the unspeakable variety of [the] Universe, turn to those phenomena that are of the nature of homogeneities of connectedness in each; and what a spectacle will unroll itself!" (6.464). As the process of Musement develops, "the Muser will naturally pass to the consideration of homogeneities and connections" between particulars and the whole and so be led to considerations of the whole itself. Finally, "in the Pure Play of Musement the idea of God's Reality will be sure sooner or later to be found an attractive fancy, which the Muser will develop in various ways" (6.464). While such a process does not "prove" the existence of God as a matter of deductive necessity, God emerges as the product of open reflection on the world in the same manner that other claims emerge without explanation in the process of a genuine inquiry. Such inquiry is an inference in the sense that the hypothesis "follows" from the situation at hand in a compelling way, but with only the limited support of a single premise—the situation itself. Students of logic set the suggestion stage of inquiry aside at their own risk, since it is here, through pure play of thought, that new ideas begin.

More formally, an inquiry begins with "some surprising phenomenon, some experience which either disappoints an expectation, or breaks in upon some habit or expectation" (Peirce, 6.469). Call this moment P. The response to the interruption or surprise is to propose some reason, cause, or situation that would give rise to the surprise. Call this prior situation S. In response to P, then, we propose: If S then P. The conditional assertion, if S then P, serves as a hypothesis, a proposal that will explain the initial situation which is characterized by surprise (or doubt or confusion). Taken together, the initial situation and the proposal, if S then P, gives reason to look for S or to try to bring S about. If S occurs and is followed by P, then we have further evidence that the hypothesis is correct. Peirce describes the inference at work this way:

> The surprising fact, [P] is observed;
> But if [S] were true, [P] would be a matter of course.
> Hence, there is reason to suspect that [S] is true. (1.189)

The generation of the hypothesis is a creative moment in which the inquirer makes a guess and, as such, is the only source of new ideas. Peirce observes that the work of inquiry is to build "a cantilever bridge" over "the chasm that yawns" between our present doubt or confusion to "the ultimate goal of science." "Yet every plank of its advance is first laid by [abduction] alone, that is to say, by the spontaneous conjectures of instinctive reason" (6.475). The process of abduction, then, begins with a situation that has no ready explanations or an indeterminate situation in which one cannot go on. In response, the inquirer makes a guess, a conjecture, a hypothesis. The fact of its emergence just at this point makes it both a creative act and reason enough to try something that may get things moving again. To track the course of an inquiry, the initial situation, P, gives rise to an assertion, if S then P, which in turn serves as justification for trying S and seeing what it brings about. The form of this process generally is

$$p$$
$$\therefore s \rightarrow p \; [\text{'if s then p'}]$$
$$\therefore \text{try s}$$

As a stage of inquiry abduction also admits of certain guiding principles that make for better and worse "guessing." The first guideline is that an inquirer must adopt a *disposition of hope*, that is, the conviction that despite evidence to the contrary, "the facts in hand admit of rationalization, and of rationalization by us." To enter an inquiry, or to find oneself in an indeterminate situation and believe firmly that there is no means of "rationalizing" or making sense of the situation will inevitably cut against one's ability to respond to the situation in a creative way. Second, insofar as the work of inquiry is the transformation of the situation—its reconstruction—it follows that abduction should occur in a context that takes account of what may come of the situation. "Now the only way to discover the principles upon which anything ought to be constructed," Peirce concludes, "is to consider what is to be done with the constructed thing after it is constructed" (7.219). Given a confidence that creative responses are possible, *making guesses in the context of future expectations* can help to ensure the relevance of ideas generated and provide a context in which the hypotheses can be tested (7.220). And third, Peirce proposes that abduction be controlled by a *commitment to "economy."* "Now economy, in general," Peirce explains, "depends upon three kinds of factors: cost; the value of the thing proposed, in itself; and its effect upon other projects" (7.220). Such considerations, he argues, provide a practical means of sorting the ideas abduction generates—not in order to eliminate more complex and "expensive" options but under the expectation that if less costly options achieve the results required to transform the situation, then such a solution would be sufficient.

The operation of abduction sets the stage for the transformation of the indeterminate situation into a determinate one. It serves as the source of hypotheses that can be tested either in terms of what is implied or in terms of experienced results. The former kind of testing, *deduction*, is the process of reasoning from hypotheses in order to determine where the hypothesis might lead if it is acted upon (including the act of holding it as true). Deduction is the process with which an inquirer traces the "necessary and probable experiential consequences" (7.203) of proposed hypotheses in order to determine the range of actions available in the context of the situation at hand and what they will lead to. This is the phase of inquiry Dewey called "testing by reason." "Abduction having suggested a theory," Peirce says, "we employ deduction to deduce from that ideal theory a promiscuous variety of consequences to the effect that if we perform certain acts, we shall find ourselves confronted with certain experiences. We then proceed to try these experiments, and if the predictions of the theory are verified, we have a proportionate confidence that the experiments that remain to be tried will confirm the theory" (8.209).

The process used to test the hypotheses and deductive conclusions in experience is *induction*. "Induction," Peirce says, "is an Argument which sets out from a hypothesis, resulting from a previous Abduction, and from virtual predictions, drawn by Deduction, of the results of possible experiments, and having performed the experiments, concludes that the hypothesis is true in the measure in which those predictions are verified, this conclusion, however, being held subject to probable modification to suit future experiments" (2.960). Once the traveler has thought about her possible means of crossing the ditch, she takes action by trying what she has determined to be the most likely way of crossing. If the attempt fails (and she hasn't fatally plunged into the ditch), she will eliminate a supposed option from the range of possibilities and try again. It is only in this last step of trying the results of abduction and deduction in experience that an inquiry can transform the initial situation into one in which the inquirer can go on. As Peirce observes, "Induction gives us the only approach to certainty concerning the real that we can have" (8.209).

Taken together, Dewey's account of inquiry and Peirce's account of the three elementary forms of reasoning provide a conceptual framework for understanding the process of inquiry. Inquiries begin in a present situation characterized by a pervasive quality of indeterminacy which is experienced as doubt, disruption, confusion, or even as an openness that approaches pure indeterminacy. In response to such a situation, the process of abduction leads to possible ways of acting and understanding. Dewey describes three different ways in which abduction can operate: problematizing the situation, suggesting possibilities, and offering hypotheses for testing. Each of these functions propose possible beliefs and actions which are then evaluated by deduction (testing by reason) and induction (testing by experience). If the inquiry leads to a transformation of the pervasive quality of the original situation into one that is determinate (unified, settled, ordered), the inquiry has been successful. In light of the success of the inquiry, it is then possible to re-present the process in a new form: an argument that, if properly constructed, can lead others through a process of inquiry that ends with the same or similar outcome or offers the possibility of critique and modification. Argument, from this perspective, is at once the outcome of inquiry and a new occasion for inquiry by others.

TRAVELING IN PEIRCE'S TERMS

When the traveler introduced by Dewey first encounters the ditch that blocks her way, she quickly decides that the problem is one of crossing the ditch. This initial decision is an abductive one in that it sets up a general hypothesis that if she manages to cross the ditch the situation will resolve and she will be able to continue her walk. While there is an unlimited array of ways to understand the indeterminacy at hand, her initial guess is not a simple given, but something "taken"—a way of ordering the situation at hand in order to transform it. Once she has problematized the situation, she begins to hypothesize about how to solve the problem she has set. The first guess, crossing the ditch by jumping, leads her to imagine making the leap, which, she decides, will not succeed. The next guess, that she might be able to use a log as a bridge, seems a likely solution on reflection, so she tries it and resumes her walk.

Dewey's overview of the stages of inquiry can be reframed in terms of Peirce's three kinds of reasoning. The processes of problematizing, suggesting and forming hypotheses are all uses of abduction. They are guesses that follow from the present situation and so are a kind of inference, but they are not certain. One can operate on a guess, but in most cases, they call for further examination or trial. The process of testing by reasoning is the use of deduction in Peirce's sense and the process of testing in overt action is a use of induction. In this light, the process of inquiry can be summarized using Peirce's terms.

Given an initial indeterminate situation,

$$\text{By abduction:} \quad p^i$$

The indeterminate situation is taken as one that *may be* a case of p (this marks the phase of setting the problem).

$$\text{By abduction:} \quad p^i$$
$$\therefore s \rightarrow p.$$

"If s then p" is a hypothesis that will transform p^i (p qualified by indeterminacy) into p (a situation that is determinately p).

$$\text{By deduction,} \quad s \rightarrow p$$
$$s$$
$$\therefore p$$

Deduction (in this very simple case) observes that, given the hypothesis, if some s occurs it should bring about some p.

$$\text{By induction,} \quad S^1 \text{ is followed by } P^1$$

An instance of s, S^1, is realized in experience and it is followed by an instance of p, in this case, P^1. The hypothesis is confirmed by a small degree and the pervasive quality of the initial system is transformed by degrees toward determinacy.

2.4 EXERCISES

2.4.1. Consider the following cases of inquiry using the resources available through published sources, print and electronic, present an overview of the inquiry mentioned. How did the inquirers problematize their situations? What hypotheses did they propose and with what results? Did the inquiry lead to a determinate situation?

> **1.** The discovery of antibiotics: Antibiotics did not exist as a treatment for bacterial infection until the 1940s. They were the result of at least two inquiries: the first by Alexander Fleming in 1928 and the second by Charles Fletcher in 1941.

> **2.** The process of musical composition: In a letter to a friend, Mozart wrote "My ideas come as they will, I don't know how, all in a stream. If I like them I keep them in my head, and people say that I often hum them over to myself. Well, if I can hold on to them, they begin to join on to one another, as if they were bits that a pastry cook should joint together in his pantry. And now my soul gets heated, and if nothing disturbs me the piece grows larger and brighter until, however long it is, it is all finished at once in my mind, so that

I can see it at a glance as if it were a pretty picture or a pleasing person. Then I don't hear the notes one after another, as they are hereafter to be played, but it is as if in my fancy they were all at once. And that is a revel (*das ist nun ein Schmaus*). While I'm inventing, it all seems to me like a fine vivid dream; but that hearing it all at once (when the invention is done), that's the best. What I have once so heard I forget not again, and perhaps this is the best gift that God has granted me." (Mozart quoted in Royce, 1896). Is this an instance of inquiry? If not, explain why not. If it is, support your conclusion by referring to the phases of inquiry illustrated within the description.

3. The Delano Grape Strike: In September 1965, an organization of farmworkers went on strike and refused to pick grapes for the table grape growers of Delano, California. Led by Cesar Chavez, who joined the strike in 1966, the farmworkers were committed to using nonviolent means, including a call to consumers to boycott grapes. In 1969, the Delano farmworkers signed a labor contract and ended the strike.

4. Turing's Test: In a 1950 paper, "Computing Machinery and Intelligence," Alan M. Turing began, "I propose to consider the question 'Can machines think?'" While the problem seems clear (to determine whether or not machines can think), Turing made several guesses about what thinking means.

5. David Brainerd's conversion: "One morning, while I was walking in a solitary place as usual, I at once saw that all my contrivances and projects to effect or procure deliverance and salvation for myself were utterly in vain; I was brought quite to a stand, as finding myself totally lost. I saw that it was forever impossible for me to do anything towards helping or delivering myself, that I had made all the pleas I ever could have made to all eternity; and that all my pleas were vain ...; and that there was no more virtue or goodness in them than there would be in my paddling with my hand in the water. ... I continued, as I remember, in this state of mind, from Friday morning till the Sabbath evening following (July 12, 1739), when I was walking again in the same solitary place. Here, in a mournful melancholy state *I was attempting to pray; but found no heart to engage in that or any other duty; my former concern, exercise, and religious affections were now gone. I thought that the Spirit of God had quite left me; but I well was not distressed; yet disconsolate, as if there was nothing in heaven or earth that could make me happy. Having been thus endeavoring to pray—though, as I thought, very stupid and senseless—*for nearly a half hour; then as I was walking in a thick grove, unspeakable glory seemed to open to the apprehension of my soul ... a new inward apprehension or view that I had of God, such as I never had before, nor anything which had the least resemblance to it ... I felt myself in a new world, and everything about me appeared with a different aspect from what it was wont to do." (David Brainerd in Edwards, 1749, 12–3).

2.4.2. Consider the following examples. Give a description of the process you would use to solve the puzzle. Explain each of the things you do. For example, you might begin by reading the problem through and then decide to draw a picture of the situation or lay out a grid. What do you do next? Try not to leave out any steps. As you write your description, indicate what phase of inquiry each of your actions represents.

1. In an effort to develop a better way to choose the most intelligent candidate as the American president, a new method is adopted by constitutional amendment. The nominees of the Republican, Democratic, and Green Parties are brought to Washington D.C. and are placed in a chamber at the Supreme Court building and seated in chairs so that each candidate can see the other two. The candidates are then shown five hats—two are black and three are white—by the Chief Justice of the Court, who then blindfolds all three candidates and places a hat on each one's head. The Chief Justice asks the candidates to remove their blindfolds and explains that whoever can correctly state the color of their own hat will be the next president. Any wrong statement, however, will lead to immediate banishment from the United States to a war zone as a member of the infantry. Guesses are therefore not a good idea. After a long pause, the Green Party candidate becomes the President of the United States. What color was the Green Party candidate's hat?

2. If you only had a 5-gallon and a 3-gallon jug, how would you measure exactly 4 gallons?

3. Consider the story of the leader of a New England Indian nation who was kidnapped in the 1730s and taken to England. King George, not renowned for his intelligence or stability, met with the leader and told him that he would be placed in a palace room with two doors. Behind one door would be a ship captain who would return him directly to North America. Behind the other was a very hungry bear, also kidnapped in America and brought to the palace. In order to help the leader choose, King George assigned two priests to the room who know the palace well. The Native American leader could ask exactly one question of one of the two priests. The King assures him that one of the priests has sworn to tell only the truth and the other to tell only lies. What question will the leader ask?

4. Decode the following string of letters. Each letter represents another and each space represents a space between words. The code is a simple substitution cipher.

LG TW AK LG TW HWJUWANWV GJ LG HWJUWANW

5. Long ago on the American prairie was a small town that became very divided over water rights for newcomers. The conflict led to a practice of public pronouncement where those in favor of water rights for newcomers always lied and those opposed to water rights for newcomers only told the truth. The first group was called the Cretans (after the famous "Liar's Paradox") and the other group was called the Puritans (thanks to their unwelcoming attitude toward the newcomers and their painful honesty). One day in the middle of the water rights controversy, three people, Abigail, Barry, and Carl, met at the intersection of the two streets that formed the town. The first person, Abigail, spoke quietly to the second person, Barry. Barry then turned to Carl and said "Abigail says that she is a Puritan and she is." Carl says in reply, "She's no Puritan. I have known her for years and she is

definitely a Cretan." How many of the three are Cretans and how many Puritans?

6. Consider the following simple jumble of letters. They can be reordered to form the names of three philosophers. Who are they?

KOEFLDLOTWLEETYTCE

7. Decode the following string of characters. In this case, the code uses a transposition scheme in which the original sentence is written on a grid with the first letter of the sentence in the top left corner and the remaining letters listed left to right in a certain number of rows. The sentence is encoded by copying the letters from the grid, top to bottom, and beginning in the upper left corner.

TNAWTERXRNDHGRHTYEPICAISCAHAEEEES

8. In a rigid legalistic future, work has come to be tightly regulated with stiff penalties for violating the rules. The government of this future state has slowly added regulations for each type of job necessary for the flourishing of the society. At one point, the government set out to regulate house painters. In order to ensure that houses are only painted by those with experience, training and the proper tools, it established the following inflexible rule for house painters: house painters shall paint the houses of all people who do not paint their own houses and only those houses. Michelle is one of the house painters subject to the rule. Shortly after the rule came into force, a serious storm damaged the exteriors of most houses and killed all of the house painters except Michelle. Though she had much work painting the damaged houses in her area, who would paint hers?

9. A very important meeting on vital economic issues ended in failure due to the fact that a consensus on a plan of action could not be reached. It was leaked to the press that this inability was due to the stubbornness of one dissenter. In a futile and very un-thought-out attempt to both protect the secrecy of the dissenter's identity and not come off as being total liars, all the participants of the meeting gave conflicting statements to the press, which ran as follows:

GEORGE   The dissenter wasn't Tom. The dissenter was Nancy.

RALPH   The dissenter was Tom. The dissenter wasn't George.

TOM   The dissenter was John. The dissenter wasn't George.

NANCY   The dissenter wasn't Ralph. The dissenter wasn't Tom.

JOHN   The dissenter was Ralph. The dissenter was Nancy.

We know that in their statements each participant told one lie and one truth. With this in mind, who was the dissenter?

10. Ed and Celia are speechwriters for a politician whose opponent is ahead in the polls. The politician is concerned that the mandate will go to her opponent. Ed and Celia prepare a brief in which they argue, first, that one vote does not make a mandate. Of course, they say, if one vote does not make a

mandate, then two votes do not either. And further, it is also the case that if two votes are not a mandate, then three votes are not a mandate. Continuing this reasoning for their entire district that consists of 34,329 voters, they conclude that the opponent will not have the mandate even if every voter supports her or him. Is this right?

2.5 ARGUMENT AS INQUIRY

In the last chapter, I proposed that an argument could be informally analyzed into a statement of its premises and conclusion. From the perspective of the theory of inquiry, this is only a start. While arguments are the products of inquiry (i.e., someone sets out to solve a problem and produces an argument as a way to test a possible solution), arguments themselves can be viewed as presentations of inquiries for purposes of critique and as resources for new inquiries that occur in the context of discussion. For purposes of critique, arguments can be seen as responses to particular indeterminate situations that display a particular process of inquiry aimed at transforming the original situation. As a re-presentation of an inquiry, new inquirers can see where the earlier effort went wrong and right, both to undo past inquiries whose results are no longer sufficient and to find resources for future inquires. In the context of new inquiries, arguments present past results for reconsideration, affirmation, rejection or change.

If arguments are understood in this way, then the insights gained about the process of inquiry also provide an approach to understanding and assessing arguments. Following the stages of inquiry, in fact, one can structure an examination of arguments in light of five questions:

1. What is the indeterminate situation from which the inquiry emerged?
2. How is the problem framed and are there alternatives?
3. Where do the postulates come from and are they warranted?
4. Do the postulates warrant the conclusion?
5. Has the indeterminate situation been transformed?

These five questions provide a template for evaluating arguments not in terms of their rhetorical character or formal validity (though these are relevant), but rather as inquiries that are grounded at one end in an experienced problematic situation and, at the other end, with the actual transformation of a situation into one that becomes a determinate starting point for new experience. A consequence of this sort of evaluation, itself an inquiry, is that the critics also take up the original argument and, by following the process implied by the argument, also have the potential to participate in inquiry and produce their own outcomes in transformed experience and new arguments.

This self-critical or self-correcting character of the process of inquiry means that the results of inquiry can be constantly under review in a way that will produce new knowledge and new experience. Dewey emphasized this point in commenting on Peirce's conception of science: "Science does not consist in any body of conclusions, but in the work of ever renewed inquiry, which never claims or permits finality but leads to ever renewed effort to learn. The sole ultimate justification of science as a method of inquiry is that if it is persisted in, it is self-correcting and tends to approach ever closer to stable common agreement of beliefs and ideas. Because science is a method of learning, not a settled body of truths, it is the hope of [human]kind" (Dewey, 1937, 484).

This evaluative approach works as well with arguments drawn from everyday problems, like that of the traveler, and from what appear to be far more abstract problems in the work of philosophy. Consider the following argument from Rudolph Carnap:

> Many words of metaphysics, now, can be shown ... to be devoid of meaning. Let us take as an example the metaphysical term 'principle' (in the sense of principle of being, not principle of knowledge or axiom). Various metaphysicians offer an answer to the question which is the (highest) 'principle of the world' (or of 'things,' of 'existence,' of 'being') .... In order to discover the meaning of the word 'principle' in this metaphysical question we must ask the metaphysician under what conditions a statement of the form 'x is the principle of y' would be true and under what conditions it would be false. In other words: if we ask for the criteria of application or for the definition of the word 'principle.' The metaphysician replies approximately as follows: 'x is the principle of y' is to mean that 'y arises out of x,' 'the being of y rests on the being of x' ... and so forth.[7] But these words are ambiguous and vague. Frequently they have a clear meaning; e.g. we say of a thing or process y that it 'arises out of' z when we observe that things or processes of kind x are frequently or invariably followed by things or processes of kind y. ... But the metaphysician tells us that he does not mean an empirically observable relationship. For in that case his metaphysical theses would be merely empirical propositions of the same kind as those of physics. The expression 'arising from' is not to mean here a relation of temporal or causal sequence. ... Yet, no criterion is specified for any other meaning. Consequently, the alleged 'metaphysical' meaning, which the word is supposed to have here in contrast to the mentioned empirical meaning, does not exist ... it remains meaningless as long as no method of verification can be described. (Carnap, 1959, 65–6)

If we consider Carnap's argument as an instance of inquiry, it can be understood at a number of points. First, since inquiries begin in an indeterminate situation, one can start by asking about what indeterminate situation gives rise to the inquiry. At times, of course, the original situation is not addressed in the text and when it is, it already appears in the form of a problem, that is, decisions have already been made about what kind of problem the situation involves. In this case, Carnap was writing in the philosophical context of the first half of the 20th century as part of movement of philosophers, often called "Logical Positivism." These philosophers, at least in part, sought to challenge the long-dominant philosophies of idealism that flourished in Germany and Austria. Part of the project of this school of philosophers was

---

[7] The "metaphysician" Carnap takes to task throughout this paper is Martin Heidegger.

to challenge the possibility of metaphysics. They held that dogmatic metaphysical claims were leading to the separation of philosophy and science and the increasing identification of philosophy with the rise of totalitarianism in Europe. In the present argument, Carnap does not relate the wider situation to his argument, but rather begins by framing a particular problem: whether metaphysical terms are meaningful.

Second, in an analysis of the argument presented, one might consider how well the argument responds to the wider situation that gave rise to the inquiry. One might raise questions about alternative approaches to problematizing the same situation or argue that the problem as framed does not lead to the expected transformation of the situation. In the case of the traveler, had she responded to her situation by trying to determine the number of plants species living in the meadow, we might conclude that her formulation of the problem missed the point. In Carnap's case, one might investigate whether or not the challenge to metaphysics was a better approach than, for example, framing the larger problem as one of dogmatism. By wondering first about the situation that gives rise to an inquiry and about the problem as it is set, it is both easier to understand how the argument begins and whether or not it is off on the "right foot."

Third, if an argument is taken as an inquiry, then the premises that are presented in the argument can also be seen as the hypotheses of an inquiry. Just as the traveler proposed to herself a number of possible ways to cross the ditch, so Carnap proposes a series of claims about the nature of meaning and metaphysical terms that provide the "grounds" for his conclusion. To approach the argument from this direction, we can first specify the premises of the argument.

> **Premise 1:** Metaphysical terms express relations, for example, the term "principle" is a relation expressed "x is a principle of y."
>
> **Premise 2:** Relational terms (such as "x is a principle of y") are either verifiable or not.
>
> **Premise 3:** If a term is verifiable, we can describe an empirical procedure of verification.
>
> **Premise 4:** If a term is meaningful, then it is verifiable.
>
> **Premise 5:** Metaphysical terms are not empirical.
>
> **Premise 6:** Metaphysical terms are not verifiable.
>
> **Conclusion:** So metaphysical terms are not meaningful.

Once the premises are set out, it is possible to examine each of the claims and assess its *origin* and *warrant*. The origin of claims can be divided into *four categories*: *hypotheses* (products of abduction), *induction* (or the products of particular experiences or sequences of experience), *deduction* (products of the relation of claims given in the argument), and *postulates* (claims that are the products of previous inquiries that are being used again in a new inquiry). The warrant of claims is tied in part to their origins. Hypotheses rarely stand on their own and call for support. Additional support can be provided through the use of deduction or induction, but the initial claim is abductive. The warrant of an inductive claim depends upon the accumulated experience that can support the claim. The traveler makes a number of inductive claims as a result of her experimentation with possible ways across the divide. Deductive warrant is usually viewed as the strongest because a claim like premise 2 would seem to hold in all cases. The postulates depend upon the inquiries that produced them.

Consider again the premises listed above in light of their possible origins.

Premise 1: Metaphysical terms express relations, for example, the term "principle" is a relation expressed "x is a principle of y."

Origin: Abduction

Comment: Assuming that Carnap is not starting with a set of stipulated definitions, he would like his readers to conclude that the situation as a whole generates the conclusion that metaphysical terms have this particular character.

Premise 2: Relational terms (such as "x is a principle of y") are either verifiable or not.

Origin: Deduction

Comment: Based on the "logical" or "formal" expectation that a term cannot be both verifiable and not verifiable.

Premise 3: If a term is verifiable, we can describe an empirical procedure of verification.

Origin: Postulate.

Comment: Although the excerpt does not present the inquiry that Carnap thinks demonstrates this premise, it is one that is supported by explicit argument, in contrast to the claim in premise 1 which is not.

Premise 4: If a term is meaningful, then it is verifiable.

Origin: Postulate.

Comment: This claim is also the product of extensive inquiry by Carnap and his colleagues. The basic idea is borrowed from David Hume who concludes that if you want to know the meaning of an idea, you should find the impression it copies, where an impression is some direct sensation or "sense datum."

Premise 5: Metaphysical terms are not empirical.

Origin: Postulate

Comment: This is the conclusion of a brief argument embedded in the larger argument. In this case, Carnap says that metaphysicians will hold that if a claim is empirical (and applies to particular things that are experienced) it cannot be metaphysical (because such principles would be equivalent to those in physics).

Premise 6: Metaphysical terms are not verifiable.

Origin: Deductive

Comment: This premise relies on our expectations about the relation of claims. In this case, since the consequent of premise 3 is denied by premise 5 in relation to metaphysical terms, we should accept the denial of the antecedent as it applies to metaphysical terms. So, premise 6 states the result: metaphysical terms are not verifiable.

Conclusion: So metaphysical terms are not meaningful.

Origin: Deductive

Comment: The conclusion can be accepted because premise 6 denies the consequent of premise 4 and so we should accept as the conclusion the denial of the antecedent of premise 4 as it applies to metaphysical terms.

In the case of each premise, the claims can be understood relative to their origins and warrant and challenged in those same terms. Note that there are no inductive premises in this case. In fact, most inquiries involve induction as well. When the traveler experimented with means to cross the ditch, she used the results of those experiments as further postulates in her inquiry. Had she offered her case for using a log to cross the ditch as an argument, she would have used her experimental results as premises to support her conclusion.

Once the argument is presented, two other things can be considered. The first (question 4, above) is whether the conclusion follows from the premises, that is, whether the argument is valid. This is a central question for logical theory and we will consider it in the next chapter. Finally (question 5, above), arguments can be evaluated in terms of their outcomes. As Dewey reminds us, inquiry that begins in an indeterminate situation is not successful until it returns to the situation that gave rise to it. It is not enough to draw conclusions without taking action. His sternest critique of professional philosophers is that their inquiries stop before returning to experience. "A first rate test of the value of any philosophy," Dewey said, is to ask "Does it end in conclusions which when they are referred back to ordinary life-experiences and their predicaments, render them more significant, more luminous to us, and make our dealing with them more fruitful?" (1925, 18).

To ask about an argument as an inquiry is also to ask about its consequences in experience. In the case of Carnap's argument, one might examine where his conclusion leads and how situations influenced by the conclusion are experienced and whether the results of the inquiry are, at least in principle, available for further use in new inquiries. One might, therefore, ask whether the postulates produced by the argument (in this case, that metaphysical terms are meaningless) can be used in other arguments. Postulates that seem well supported in one argument might turn out to lead to further conclusions that are challenged by other postulates and further inquiry.

To summarize, one can, first, examine an argument relative to its initiating situation and, second, how it has been framed as a problem; third, examine the premises at hand relative to their origin and warrant; fourth, consider whether the conclusions follow from the premises; and fifth, examine whether and how the argument as inquiry returns to the initial indeterminate situation and transforms it and how its results carry into new situations.

2.6 EXERCISES

Consider the following arguments as inquiries. Following the model given in 2.5, summarize the premises and conclusion(s) and give the origin of each premise in the process of inquiry.

1. "[T]he argument [for democracy] must be continually restated and emphasized. We must remember that if the theory of democracy is correct, the right to vote is not merely a privilege, not simply a method of meeting the needs of a particular group, and least of all a matter of recognized want or desire. Democracy is a method of realizing the broadest measure of justice to all human beings. The world has, in the past, attempted various methods of attaining this end. ... [But as] soon as a nation discovers that it holds in the heads and hearts of its individual citizens the vast mine of knowledge, out of which it may build a just government, then more and more it calls those citizens to select their rulers and to judge the justice of their acts.

"... The temptation is to ask only for the wisdom of citizens of a certain grade or those of recognized worth. Continually some classes are tacitly or expressly excluded. Thus women have been excluded from modern democracy because of the persistent theory of female subjection and because it was argued that their husbands or other male folks would look to their interests. Now, manifestly, most husbands, fathers, and brothers will, so far as they know how or as they realize women's needs, look after them. But remember the foundation of the argument,—that in the last analysis only the sufferer knows his sufferings and that no state can be strong which excludes from its expressed wisdom the knowledge possessed by mothers, wives, and daughters. We have but to view the unsatisfactory relations of the sexes the world over and the problem of children to realize how desperately we need this excluded wisdom.

"The same arguments apply to other excluded groups: if a race, like the Negro race, is excluded, then so far as that race is a part of the economic and social organization of the land, the feeling and the experience of that race are absolutely necessary to the realization of the broadest justice for all citizens. Or if the 'submerged tenth' be excluded, then again, there is lost from the world an experience of untold value, and they must be raised rapidly to a place where they can speak for themselves. In the same way and for the same reason children must be educated, insanity prevented, and only those put under the guardianship of others who can in no way be trained to speak for themselves.

"The real argument for democracy is, then, that in the people we have the source of that endless life and unbounded wisdom which the rulers of men must have." (W. E. B. Du Bois, 1920)

2. "The Meditation of yesterday filled my mind with so many doubts that it is no longer in my power to forget them. ... I shall nevertheless make an effort and follow anew the same path as that on which I yesterday entered, i.e., I shall proceed by setting aside all that in which the least doubt could be supposed to exist, just as if I had discovered that it was absolutely false; and I shall ever follow in this road until I have met with something which is certain, or at least, if I can do nothing else, until I have learned for certain that there is nothing in the world that is certain. ...

"I suppose, then, that all the things that I see are false; I persuade myself that nothing has ever existed of all that my fallacious memory represents to me. I consider that I possess no senses; I imagine that body, figure, extension, movement and place are but the fictions of my mind. What, then, can be esteemed as true? Perhaps nothing at all, unless that there is nothing in the world that is certain.

"But how can I know there is not something different from those things that I have just considered, of which one cannot have the slightest doubt? Is there not some God, or some other being by whatever name we call it, who puts these reflections into my mind? That is not necessary, for is it not possible that I am capable of producing them myself? I myself, am I not at least something? But I have already denied that I had senses and body. Yet I hesitate, for what follows from that? Am I so dependent on body and senses that I cannot exist without these? But I was persuaded that there was nothing in all the world, that there was no heaven, no earth, that there were no minds, nor any bodies: was I not then likewise persuaded that I did not exist? Not at all; of a surety I myself did exist since I persuaded myself of something [or merely because I thought of something]. But there is some deceiver or other, very powerful and very cunning, who ever employs his ingenuity in deceiving me. Then without doubt I exist also if he deceives me, and let him deceive me as much as he will, he can never cause me to be nothing so long as I think that I am something. So that after having reflected well and carefully examined all things, we must come to the definite conclusion that this proposition: I am, I exist, is necessarily true each time that I pronounce it, or that I mentally conceive it." (Descartes, 1911)

3. Socrates said: "When I heard the answer, I said to myself, What can the god mean? and what is the interpretation of his riddle? for I know that I have no wisdom, small or great. What then can he mean when he says that I am the wisest of men? And yet he is a god, and cannot lie; that would be against his nature. After long consideration, I thought of a method of trying the question. I reflected that if I could only find a man wiser than myself, then I might go to the god with a refutation in my hand. I should say to him, 'Here is a man who is wiser than I am; but you said that I was the wisest.' Accordingly I went to one who had the reputation of wisdom, and observed him—his name I need not mention; he was a politician whom I selected for examination—and the result was as follows: When I began to talk with him, I could not help thinking that he was not really wise, although he was thought wise by many, and still wiser by himself; and thereupon I tried to explain to him that he thought himself wise, but was not really wise; and the consequence was that he hated me, and his enmity was shared by several who were present and heard me. So I left him, saying to myself, as I went away: Well, although I do not suppose that either of us knows anything really beautiful and good, I am better off than he is,—for he

knows nothing, and thinks that he knows; I neither know nor think that I know. In this latter particular, then, I seem to have slightly the advantage of him ....

"Then I went to one man after another, being not unconscious of the enmity which I provoked, and I lamented and feared this: but necessity was laid upon me,—the word of God, I thought, ought to be considered first. And I said to myself, Go I must to all who appear to know, and find out the meaning of the oracle. And I swear to you, Athenians, by the dog I swear!—for I must tell you the truth—the result of my mission was just this: I found that the men most in repute were all but the most foolish; and that others less esteemed were really wiser and better. ...

"This inquisition has led to my having many enemies of the worst and most dangerous kind, and has given occasion also to many calumnies. And I am called wise, for my hearers always imagine that I myself possess the wisdom which I find wanting in others: but the truth is, O men of Athens, that God only is wise; and by his answer he intends to show that the wisdom of men is worth little or nothing; he is not speaking of Socrates, he is only using my name by way of illustration, as if he said, He, O men, is the wisest, who, like Socrates, knows that his wisdom is in truth worth nothing. And so I go about the world, obedient to the god, and search and make enquiry into the wisdom of any one, whether citizen or stranger, who appears to be wise; and if he is not wise, then in vindication of the oracle I show him that he is not wise; and my occupation quite absorbs me, and I have no time to give either to any public matter of interest or to any concern of my own, but I am in utter poverty by reason of my devotion to the god." (Plato, 1909, 7–9)

4. "What, then, is the rightful limit to the sovereignty of the individual over himself? Where does the authority of society begin? How much of human life should be assigned to individuality, and how much to society?

"Each will receive its proper share, if each has that which more particularly concerns it. To individuality should belong the part of life in which it is chiefly the individual that is interested; to society, the part which chiefly interests society.

"Though society is not founded on a contract, and though no good purpose is answered by inventing a contract in order to deduce social obligations from it, every one who receives the protection of society owes a return for the benefit, and the fact of living in society renders it indispensable that each should be bound to observe a certain line of conduct towards the rest. This conduct consists, first, in not injuring the interests of one another; or rather certain interests, which, either by express legal provision or by tacit understanding, ought to be considered as rights; and secondly, in each person's bearing his share (to be fixed on some equitable principle) of the labors and sacrifices incurred for defending the society or its members from injury and molestation. These conditions society is justified in enforcing, at all costs to those who endeavor to withhold fulfillment. Nor is this all that society may do. The acts of an individual may be hurtful to others, or wanting in due consideration for their welfare, without going to the length of violating any of their constituted rights. The offender may then be justly punished by opinion, though not by law. As soon as any part of a person's conduct affects prejudicially the interests of others, society has jurisdiction over it, and the question whether the general welfare will or will not be promoted by interfering with it, becomes open to discussion. But there is no room for entertaining any such question when a person's conduct affects the interests of no persons besides himself, or needs not affect them unless they like (all the persons concerned being of full age, and the ordinary amount of understanding). In all such cases, there should be perfect freedom, legal and social, to do the action and stand the consequences." (Mill, 1863, 144–6)

5. "To account for, and excuse the tyranny of man, many ingenious arguments have been brought forward to prove, that the two sexes, in the acquirement of virtue, ought to aim at attaining a very different character: or, to speak explicitly, women are not allowed to have sufficient strength of mind to acquire what really deserves the name of virtue. Yet it should seem, allowing them to have souls, that there is but one way appointed by Providence to lead mankind to either virtue or happiness.

"If then women are not a swarm of ephemeron triflers, why should they be kept in ignorance under the specious name of innocence? Men complain, and with reason, of the follies and caprices of our sex, when they do not keenly satirize our headstrong passions and groveling vices. Behold, I should answer, the natural effect of ignorance! The mind will ever be unstable that has only prejudices to rest on, and the current will run with destructive fury when there are no barriers to break its force. Women are told from their infancy, and taught by the example of their mothers, that a little knowledge of human weakness, justly termed cunning, softness of temper, outward obedience, and a scrupulous attention to a puerile kind of propriety, will obtain for them the protection of man; and should they be beautiful, every thing else is needless, for, at least, twenty years of their lives. ...

"How grossly do they insult us who thus advise us only to render ourselves gentle, domestic brutes! For instance, the winning softness so warmly, and frequently, recommended, that governs by obeying. What childish expressions, and how insignificant is the being—can it be an immortal one? who will condescend to govern by such sinister methods! ... Men, indeed, appear to me to act in a very unphilosophical manner when they try to secure the good conduct of women by attempting to keep them always in a state of childhood. Rousseau was more consistent when he wished to stop the progress of reason in both sexes, for if men eat of the tree of knowledge, women will come in for a taste; but, from the imperfect cultivation which their understandings now receive, they only attain a knowledge of evil.

"Children, I grant, should be innocent; but when the epithet is applied to men, or women, it is but a civil term for weakness. For if it be allowed that women were destined by Providence to acquire human virtues, and by the exercise of their understandings, that stability of character which is the firmest ground to rest our future hopes upon, they must be permitted to turn to the fountain of light, and not forced to shape their course by the twinkling of a mere satellite." (Wollstonecraft, 1891, 49–50)

CHAPTER 3

COMMUNICATIVE ACTION

3.1   STRATEGIC AND COMMUNICATIVE ACTION

In the last chapter, we considered inquiry and arguments as a form of inquiry. From this angle, arguments are a presentation of a response to an indeterminate situation where the premises mark the stages of inquiry and the resources used. Arguments can then serve as a resource for the reconsideration of past results and as a resource for new inquiries in response to new indeterminate situations. The combined function of arguments is to be taken up into new discussions and made part of what Jürgen Habermas, a contemporary philosopher of language and social theory, called the "community of investigation." Here, the results of particular inquiries—perhaps conducted in isolation—are brought into a larger conversation and tested anew.

Habermas, writing about Peirce's theory of inquiry, concludes "Deduction, induction, and abduction establish relations between statements that are, in principle, [part of a monologue]. It is possible to think in syllogisms, but not to conduct a dialogue in them. I can use a syllogistic reasoning to yield arguments for a discussion, but I cannot argue syllogistically with another" (1971, 137). Taken up into a dialogue, arguments become part of a process that involves more than one inquirer "who know and recognize each other as unmistakable individuals" (1971, 137). In this context, the presentation of an argument itself becomes a kind of action that can be done in ways that foster new inquiries and the transformation of new indeterminate situations, or can be done in ways that undermine inquiry and coerce results.

The use of language in speaking or writing is, of course, an activity. Different linguistic activities can be carried out for different purposes. To shout "Look out!" at a pedestrian crossing a street without noticing a fast approaching car has the purpose of warning the pedestrian. To mutter to one's self while completing a dull homework assignment might have no greater purpose than to vocalize one's dislike of the task. If we confine the issue to language acts involving assertions, we can see that there are still at least two broad kinds of purpose, both of which can be understood as a kind of "interaction [that is] the solution to the problem of how the action plans of several actors can be coordinated in such a way that the action of Alter [another] can be connected up with those of Ego [one's self]" (Habermas, 1998, 221).

Consider cases in which a salesperson is explaining the virtues of her product to a prospective buyer. By stating the benefits, the salesperson is not attempting to engage in a conversation over the merits of the product in a way that would lead to new ideas about the product or its disadvantages relative to something else. Rather, the assertions made aim at causing the prospective buyer to buy the product. Or consider cases in which politicians "make a case" for war or tax cuts; despite protests that such speech aims at "open debate," in the course of the interaction it often appears that the speaker is more interested in moving the audience to actively support the proposal (or at least not to actively oppose it). In these cases, the use of language is aimed at causing the audience to do (or not do) something through the "transmission of information" in which the effect of language "remains dependent on the influence—functioning via nonlinguistic activities—exerted by actors on the action situation and on each other" (1998, 221). Habermas views this use of language as *strategic action*.

In contrast to strategic action, language is also used as "a source of social integration" in which "the consensus achieving force of linguistic processes of *reaching understanding* …—that is, the binding and bonding energies of *language itself*—becomes effective for the coordination of actions" (1998, 221). The second use of language is called *communicative action*. In this case, the use of language aims at seeking agreement, not simply achieving some particular, predetermined end. Agreement is not a kind of attitude that can be imposed; it must be arrived at by each listener. In the case of strategic action, information is transmitted in language to bring about the coordination of diverse actors in light of goals or expectations decided in advance. In contrast, communication aims at promoting understanding so that, through understanding, diverse actors will choose to coordinate their actions in ways that may or may not have been anticipated when the communication began. When engaged in a process of communication, actors come to understand each other's plans and interests, cooperatively define the situation, and then attempt to harmonize their aims in order to take collective action. Communicative action, Habermas concludes, is different from strategic action in that it relies on "rationally motivated feats of understanding" by all participants, not just on the plans by a few imposed on others.

Habermas observes that there are two sorts of strategic speech acts. In the first, the speaker claims to be engaged in communicative action, but is primarily interested in causing a particular outcome. The others involved take the speaker at her or his word. The bias on the part of the participants is to expect communicative action; the speaker can use this expectation to mask a strategic effort. What such action demonstrates is that there is an "independent logic underlying linguistic communication—a logic that is effective for coordination only to the extent that it subjects the purposive activities of the actors to certain constraints" (1998, 224).

Habermas also recognizes a second form of strategic action in which communicative practices become limited to an exchange of information only. When someone shouts to warn another about a danger, the process of agreement is replaced by language meant to cause certain actions regardless of consent or understanding. From this perspective, strategic action can be seen in a range of linguistic acts that include deception, coercion, and threat. In any of these cases, the goal of communication is set aside in favor of bringing about results regardless of agreement.

In general, strategic actions are taken in light of goals that are determined independently of the means of achieving the goals. In other words, the goal of making someone do what you want (buy a product, vote for you, sell a valued gift) are chosen without regard to what it might take to achieve the goal. The speech acts that (you hope) will bring about the result are done in service of the goal and are viewed as causes of the result. In order to carry out the process, the speaker must view the other participants (the audience for the speech act) as objects who can be caused to act in accordance with the speaker's goals. Exchanges (questions, concerns, rejections) are not seen as efforts at communication, but rather as indicators of what further speech acts might be required to bring about the goal in question.

In contrast, communicative action sets out to foster understanding and agreement. Here the goal of the speech act—"illocutionary goals"—cannot be separated from the means of achieving the goal. What you say to bring about understanding must be such that it makes way for agreement. The result is not caused, but rather occasioned in the sense that even as you present an argument, the audience is viewed as part of the exchange and must choose to agree to the conclusion offered. Agreement cannot be imposed but requires a decision to take the action in light of the goals the action is to realize. If someone acts because they are threatened or fooled, or because they are promised some reward independent of the results of the action, their participation does not count as agreement. Communication is then a process that seeks to foster a willful decision on the part of agents involved in a situation to take or not to take an action. The decision to act or not is the key result of communicative action and it is the central logical operation as we shall see later.

Some might argue that human communication is best seen as originating in strategic action, perhaps in order to foster individual survival from the perspective of some theory of human evolution or as part of a conception of the nature of individuals of the sort proposed by Thomas Hobbes and the liberal tradition that followed him. Habermas argues instead that human communication originates in the process of communication itself. Individuals have their origins as individuals in conversation with others and develop their priorities and interests in the context of interaction with others. Liberal individualism, on this view, is not the prior state of affairs, but a product of a particular history of communication in which participants learn to prioritize their own interests and view themselves as fundamentally independent of others. Despite the conclusions of liberal individualism, however, even these individuals require a communal or communicative starting point. Similarly, when one considers the process of strategic speech action, it is clear that much of it relies on already established habits of communication. It is not possible to lie to someone, for instance, if both participants do not recognize that speech is first a matter of communication in which things are presented in a way that aims at agreement. A lie occurs when someone makes a statement that is taken as if it had the goal of communication, even though the speaker intends some other strategic outcome.

More important to the primacy of communicative action, however, is the recognition that the strategic use of language is one which proposes goals that are outside the context of an exchange between speaker and audience. Since the goals of strategic speech are not found within the exchange, conversation becomes little more than an unbroken series of causes and effects. New ideas and critical challenges can only amount to emergent obstacles in the already established effort to bring about some particular goal (buying a used car, having sex, going to war). The world of strategic speech acts is a world without shared reflection and without new ideas.

Communicative speech acts on the other hand make room for new ideas and goals because obstacles in the course of communication mark new opportunities. Since the audience is part of the exchange, objections, counterarguments, and new proposals are all part of the process and lead to a richer array of outcomes. None of this is to suggest that strategic action is only evil, but rather to suggest that strategic speech acts must be understood as a way to use language which is dependent upon a prior use of language that involves others and makes innovation and agreement possible. It is only in the case of communicative action that shared efforts of the participants "step out of the egocentricity of a purposive rational orientation toward their own respective success and to surrender themselves to the public criteria of communicative rationality" (Habermas, 1998, 233).

3.2   EXERCISES

3.2.1. Write two short dialogues illustrating an example of strategic action and an example of communicative action. Each dialogue should involve two speakers and five speech acts. Conclude with a paragraph comparing the two dialogues.

3.2.2. Consider the following examples. In each case, decide whether it is an example of strategic or communicative action and explain your conclusion.

1. **Jane Addams on War.**

   In 1915, shortly after the beginning of World War One, Jane Addams, a social activist from Chicago and co-founder of Hull House, travelled to Europe with a group of women peace advocates. The women met with the leadership of most of the countries that declared war and also met with soldiers, medical personnel, and the mothers of soldiers. On her return from Europe, Addams gave a speech at Carnegie Hall in New York City. The following excerpt is from the conclusion of her address.

   "The old notion that you can drive a belief into a man at the point of a bayonet is in force once more. It is quite as foolish to think that if militarism is an idea and an ideal, it can be changed and crushed by counter-militarism or by a bayonet charge. And the young men in these various countries say of the bayonet charges: 'That is what we cannot not think of.' We heard in all countries similar statements in regard to the necessity of the use of stimulants before men would engage in bayonet charges—that they have a regular formula in Germany, that they give them rum in England and absinthe in France; that they all have to give them 'dope' before the bayonet charge is possible. Well, now, think of that.

   "No one knows who is responsible for the war; all the warring nations are responsible, and they indict themselves. But in the end human nature must reassert itself. The old elements of human understanding and human kindliness among them must come to the fore, and then it may well be that they will reproach the neutral nations and will say: 'What was the matter with the rest of the world that you kept quiet while this horrible thing was happening, and our men for a moment had lost their senses in this fanaticism of national feeling all over Europe?'" (Addams, 1915, 359)

2. **St. Anselm's Conversation (Argument) with and About God**

   "And so, Lord, do thou, who dost give understanding to faith, give me, so far as thou knowest it to be profitable, to understand that thou art as we believe; and that thou art that which we believe. And, indeed, we believe that thou art a being than which nothing greater can be conceived. Or is there no such nature, since the fool hath said in his heart, there is no God? But, at any rate, this very fool, when he hears of this being of which I speak—a being than which nothing greater can be conceived—understands what he hears, and what he understands is in his understanding; although he does not understand it to exist.

   "For, it is one thing for an object to be in the understanding, and another to understand that the object exists. When a painter first conceives of what he will afterwards perform, he has it in his understanding, but he does not yet understand it to be, because he has not yet performed it. But after he has made the painting, he both has it in his understanding, and he understands that it exists, because he has made it. Hence, even the fool is convinced that something exists in the understanding, at least, than which nothing greater can be conceived. For, when he hears of this, he understands it. And whatever is understood, exists in the understanding. And assuredly that, than which nothing greater can be conceived, cannot exist in the understanding alone.

For, suppose it exists in the understanding alone: then it can be conceived to exist in reality; which is greater.

"Therefore, if that, than which nothing greater can be conceived, exists in the understanding alone, the very being, than which nothing greater can be conceived, is one, than which a greater can be conceived. But obviously this is impossible. Hence, there is no doubt that there exists a being, than which nothing greater can be conceived, and it exists both in the understanding and in reality." (Saint Anselm, 1903)

3. George W. Bush Justifies Use of Force on Iraq

"Some citizens wonder, after 11 years of living with this problem, why do we need to confront it now. And there's a reason. We have experienced the horror of Sept. 11. We have seen that those who hate America are willing to crash airplanes into buildings full of innocent people. Our enemies would be no less willing, in fact they would be eager, to use biological or chemical or nuclear weapons. Knowing these realities, America must not ignore the threat gathering against us. Facing clear evidence of peril, we cannot wait for the final proof, the smoking gun, that could come in the form of a mushroom cloud. ... Some believe we can address this danger by simply resuming the old approach to inspections and applying diplomatic and economic pressure. Yet this is precisely what the world has tried to do since 1991. ... After 11 years during which we've tried containment, sanctions, inspections, even selected military action, the end result is that Saddam Hussein still has chemical and biological weapons and is increasing his capabilities to make more. And he is moving ever closer to developing a nuclear weapon. ... The time for denying, deceiving and delaying has come to an end. Saddam Hussein must disarm himself or, for the sake of peace, we will lead a coalition to disarm him. ... There is no easy or risk-free course of action. Some have argued we should wait, and that's an option. In my view it's the riskiest of all options because the longer we wait, the stronger and bolder Saddam Hussein will become. ... As Americans we want peace, we work and sacrifice for peace. But there can be no peace if our security depends on the will and whims of a ruthless and aggressive dictator. ... Failure to act would embolden other tyrants, allow terrorists access to new weapons and new resources, and make blackmail a permanent feature of world events. The United Nations would betray the purpose of its founding and prove irrelevant to the problems of our time. And through its inaction, the United States would resign itself to a future of fear. ... Later this week the United States Congress will vote on this matter. I have asked Congress to authorize the use of America's military, if it proves necessary, to enforce U.N. Security Council demands." (Bush, 2002)

4. Hume: Philo and Cleanthes

"[Philo states the following to Cleanthes] ... For, as the supreme Being is allowed to be absolutely and entirely perfect, whatever differs most from him departs the farthest from the supreme standard of rectitude and perfection. ... These, CLEANTHES, are my unfeigned sentiments on this subject; and these sentiments, you know, I have ever cherished and maintained. But in proportion to my veneration for true religion, is my abhorrence of vulgar superstitions; and I indulge a peculiar pleasure, I confess, in pushing such principles, sometimes into absurdity, sometimes into impiety. And you are sensible, that all bigots, notwithstanding their great aversion to the latter above the former, are commonly equally guilty of both.

"My inclination, replied CLEANTHES, lies, I own, a contrary way. Religion, however corrupted, is still better than no religion at all. The doctrine of a future state is so strong and necessary a security to morals, that we never ought to abandon or neglect it. For if finite and temporary rewards and punishments have so great an effect, as we daily find: How much greater must be expected from such as are infinite and eternal?

"How happens it then, said PHILO, if vulgar superstition be so salutary to society, that all history abounds so much with accounts of its pernicious consequences on public affairs? Factions, civil wars, persecutions, subversions of government, oppression, slavery; these are the dismal consequences which always attend its prevalency over the minds of men. If the religious spirit be ever mentioned in any historical narration, we are sure to meet afterwards with a detail of the miseries which attend it. And no period of time can be happier or more prosperous, than those in which it is never regarded, or heard of.

"The reason of this observation, replied CLEANTHES, is obvious. The proper office of religion is to regulate the heart of men, humanize their conduct, infuse the spirit of temperance, order, and obedience; and as its operation is silent, and only enforces the motives of morality and justice, it is in danger of being overlooked, and confounded with these other motives. When it distinguishes itself, and acts as a separate principle over men, it has departed from its proper sphere, and has become only a cover to faction and ambition." (Hume, 1779, 241–44)

5. George Berkeley: Hylas and Philonous

PHILONOUS   Those bodies therefore, upon whose application to our own, we perceive a moderate degree of heat, must be concluded to have a moderate degree of heat or warmth in them: and those, upon whose application we feel a like degree of cold, must be thought to have cold in them.

HYLAS   They must.

PHILONOUS   Can any doctrine be true that necessarily leads a man into an absurdity?

HYLAS   Without doubt it cannot.

PHILONOUS   Is it not an absurdity to think that the same thing should be at the same time both cold and warm?

HYLAS   It is.

PHILONOUS   Suppose now one of your hands hot, and the other cold, and that they are both at once put into the same vessel of water, in an intermediate state; will not the water seem cold to one hand, and warm to the other?

HYLAS   It will.

PHILONOUS   Ought we not therefore by your principles to conclude, it is really both cold and warm at the same time, that is, according to your own concession, to believe an absurdity.

HYLAS   I confess it seems so.

PHILONOUS   Consequently, the principles themselves are false, since you have granted that no true principle leads to an absurdity.

HYLAS   But after all, can anything be more absurd than to say, there is no heat in the fire? (Berkeley, 1776, 12–5)

3.3   LIFEWORLDS

At the center of Habermas's theory of communicative action is the recognition that communication occurs in a context that provides both the resources for and constraints on the use of language. This same context for language use also provides the intellectual and practical tools necessary for responding to indeterminate situations and for constructing arguments that lead to belief. Following Edmund Husserl in *The Crisis of European Sciences*, Habermas calls the contexts for language use "lifeworlds." "The explicit feats of communication achieved by communicative actors," Habermas explains, "take place within the horizon of shared, unproblematic convictions; the disquiet that arises through experience and critique crashes against the … broad and imperturbable rock projecting out from the deep of agreed-upon interpretative patterns, loyalties, and proficiencies" (1998, 237). Lifeworlds, then, are part of the continuous character of ordinary experience. Most of what we say in our conversations with others is taken as unproblematic and are rarely subject to criticism. Such language emerges from a complex of habits that provide reliable resources for getting around. When problems arise, these habits provide a means of addressing them, even though, at times, well-established habits are also called into question and rejected or revised in the course of inquiry. To the extent that these habits are unchallenged, they operate as "the certainties of lifeworlds." The disruption that gives rise to inquiry and argument happens in part because the "certainties" of the lifeworld are blocked and prevented from functioning smoothly. A situation becomes indeterminate and aspects of the lifeworld are called into question even as other aspects come forward to frame the problem and seek to restore or transform the situation.

Indeterminate situations call into question otherwise unproblematic claims of everyday communication and we respond by seeking ways of rendering these claims plausible by calling on what Habermas called "*foreground knowledge*" (1998, 240). Such knowledge is of two sorts: "a situation specific horizontal knowledge" and "a topic-dependent contextual knowledge" (1998, 241). Horizontal knowledge regards "the perceived environment" that is "embedded in concentrically arranged spatiotemporal horizons that are not perceived" but which constitute "the center of the speech situation." This aspect of the lifeworld provides organizing relations that are presumed by speakers as they develop the postulates of inquiry and the premises of argument. To return to our earlier example, should someone ask our traveler why she has stopped her walk across the meadow, she would point out the ditch. More likely, a companion of the traveler would not ask about the ditch since it is part of the situation specific knowledge, but instead will presume it as she raises questions about how to proceed. Knowledge of this sort forms the "horizon" of the communicative situation.

The second sort of foreground knowledge, "topic-dependent contextual" knowledge, amounts to what "a speaker can presuppose within the framework of a common language, the same culture, similar schools, and so forth—that is, within the framework of a common milieu or horizon of subjective experience" (1998, 241). Such knowledge amounts to commonly accepted expectations about situations and how problems are solved in the communicative effort. Should others in the situation signal confusion about some claim I have made or my approach to framing the problem, I will appeal to this contextual knowledge on one hand and situation specific horizontal knowledge on the other to provide "information and reasons" that can be mobilized in support.

In addition to these aspects of a lifeworld distinguished by their close relation to a present communicative situation, there is another kind of knowledge that extends well beyond (or below) the present situation and serves as a body of knowledge and practices upon which the possibility of communicative action rests. While one can thematize the foreground knowledge of a situation as horizontal and contextual knowledge, the shared knowledge that underlies communication and makes it possible is better understood as "*Deepseated background knowledge* ... [that] cannot intentionally be brought to consciousness in the same way as is possible with [the foreground and] ... it forms a deep stratum of unthematized knowledge in which the situation-related horizontal knowledge and topic-dependent contextual knowledge ... have their roots" (1998, 242).

Background knowledge is distinguished from foreground knowledge by three characteristics. First, background knowledge has the character of "immediate certainty," that is, it is knowledge that operates without reflective pause and so has little chance to become problematic and the object of inquiry. Instead, it retains a kind of absolute character until, all at once, it collapses or is set aside. For example, the certainty, in Europe in the 15th century, that there were neither landmasses nor people in the Western Hemisphere, unchallenged for more than a thousand years, collapsed in the first decades of the 16th century. Supposed truths of science and beliefs about the nature of human beings have undergone rapid and dramatic transformation as well.

Second, background knowledge has the character of "totalizing power" (1998, 244). From this perspective, the background knowledge within a lifeworld "constitutes a totality" that includes a core of knowledge and a vague boundary that contracts rather than allowing new knowledge to cross and destabilize the whole. As new knowledge becomes part of the communicative situation, it finds a place at the "edges" of the world. In this way, the core commitments of the lifeworld remain unchanged and central to problematizing situations and developing conclusions. New knowledge is consequently peripheral, acknowledged in a sense but powerless in most communicative situations. In the 15th century, as new knowledge developed through the European encounters with America and its people, it was added to the background of the European lifeworld. Rather than overthrowing (or transcending) the established core of beliefs about the history of the Western Hemisphere, the new knowledge and perhaps those people whom it regarded were added to the periphery, leaving such knowledge and life marginalized.

The third characteristic of background knowledge is its "holism" (1998, 246). While one might be able to identify components of the lifeworld that are instances of certainty and totalization, the apparent "transparency" of these characteristics the lifeworld is "render[ed] impenetrable" by the network of connections that bind established knowledge to the perspectives needed for critique. "The lifeworld," Habermas concludes, "may be described as a 'thicket'" (1998, 244) that makes challenging the core commitments of a lifeworld an extremely difficult task. Taken together, these three characteristics of background knowledge provide for what is certain in our experience as well as the framework against which surprises emerge and give rise to inquiry.

Consider the case of a labor negotiation. The workers and management share a situation in which the horizontal knowledge includes a recognition of the shared workplace, something about what is produced, the ways in which the workers and management might respond to the conflict and so on. The contextual knowledge of the situation includes the implications of the various possible responses, how labor conflicts have worked out in other situations, expectations about those who advocate for one outcome or another and so on. The background is the wide cultural context in which discussions of labor conflict make sense. In this case, the background is a culture in which workers and managers take different sides, where workers can complain, where products are made and sold, where regulation plays a certain role, and so on. Communication, should it occur, relies on the vast background of a common cultural context and the ability of the participants to understand the shared aspects of the present circumstances and to view results in a way that can lead to collective action to resolve the original, troubled situation.

Access to a lifeworld emerges in the activity of communication itself: "The lifeworld as a whole comes into view only when we, as it were, stand behind the back of the actor and view communicative action as an element of a circular process in which the actor no longer appears as the initiator but rather as the product of the traditions within which she is situated, of solidarity groups to which she belongs, of socialization and learning processes to which she is subjected" (1998, 246). Once we obtain this sort of viewpoint by reflecting on the process of communication, the lifeworld can be seen as operating in and through communication in three processes: reaching understanding, coordinating action, and of socialization (1998, 247). In order to reach understanding, participants must share at least some knowledge and communicative resources. The array of possibilities for action is constrained by these resources that include paradigms of interpretation and the values in terms of which judgments are made. What actions follow from coordination within a lifeworld are inherently "conservative" in that they conserve the commitments and structures of the lifeworld. This conserving coordination is also a project of socializing the participants by affirming their common background and their place in the activities that make up the lifeworld. Social roles and practices and aspects of identity emerge in the process of communication and are reinforced.

The background knowledge that frames communication should be understood as part of the means by which the lifeworld "survives" and reproduces its components. These components are, first, *culture*, that is, "stock knowledge from which participants in communication ... supply themselves with interpretations" (1998, 249). In this sense, culture is embodied in action as well as the symbolic context in which action is considered, that is, the accepted ways of understanding aspects of experience as well as the material symbolic forms with which we interact such as books, advertising, clothes, memorials, and electronic media.

The second component, *society*, is "embodied in institutional orders, legal norms, or in webs of normatively regulated practices and customs" which "regulate affiliations to social groups and safeguard solidarity" (1998, 249). Practices of obeying traffic signals, buying and selling at the local market, participating in celebrations, preparing meals, deferring to elders, and so on are all aspects of the social background. Social learning, both in formal education and in everyday experience, helps to reinforce these ordering practices and transfer them to new generations.

Third, a lifeworld includes *personality structures* which are "embodied—in a literal sense—in the substratum of human organisms" (1998, 250), which provide "all motives and competencies that enable a subject to speak and act and thereby to secure her own identity" (1998, 248). Through these components, the lifeworld also selects who may participate in communicative action and how, as well as whose interests count and how they are taken into account in the process of presenting and justifying arguments that determine priorities and coordinate social action.

Lifeworlds, then, provide a ground for strategic action, in which persons carry out attempts to realize goals framed by culture and society, and communicative action, in which persons interact in an effort to examine and come to agreements about aspects of the lifeworld which have been disrupted. It is in the context of a lifeworld that we can call into question knowledge and structures that are related to problematic situations and also locate the resources relevant to carrying out inquiry and argumentation.

3.4 EXERCISES

3.4.1. Reconsider the examples in 3.2.2 In each case, determine some of the horizontal, contextual, and background information one would use to understand the argument.

3.4.2. Locate one or more arguments presented in some public forum, electronic or print. Determine some of the horizontal, contextual, and background information in the argument. What are some ways that culture, society, and personality play a role in these arguments?

3.5   VALIDITY

In Chapter Two, we argued that logic is concerned with the process of inquiry and argument as a form of inquiry. Successful arguments, in this context, are ones that are both valid and sound. A *valid argument*, recall, is an argument in which, if the premises are true (by whatever standard of truth we accept), then the conclusion must also be true. A *sound argument* is one which is valid and which has premises that are true. Habermas has extended our discussion of argument by considering communicative action as argument and not simply as a matter of making assertions (true or false) and seeing the consequences come to pass without further consideration. Agreement cannot be imposed except by a cooperative interaction. Language, in the context of argument, is a means of reaching understanding by providing the participants with an opportunity to hear each other's plans, questions, and justifications and then carry out action within their shared lifeworld. The possibility of cooperation and agreement in argument turns on the meaning of language and the validity of its connections.

Language meaning, Habermas observes, has traditionally been understood in three ways, as intentional, formal, and as a matter of action or use. Intentional meaning, that is, *intentional semantics*, focuses on our recognition that language meaning is connected to what speakers intend. If you want to understand my assertion that the Pacific coast has an average water temperature of 52 degrees, then you need to understand that I intend to speak about the Pacific coast of the United States (even if I accidentally say "Atlantic"). An assertion, one can say, properly means what the speaker intends it to mean and so understanding depends upon getting hold of what is in "the mind" of the speaker.

Truth semantics, on the other hand, focuses on the aspect of language meaning that connects assertions to the things that they refer to. If I say "The Atlantic coast has an average water temperature of 52 degrees," then the meaning of the assertion is some state of affairs in the world. If I have misspoken—I meant the Pacific coast!—it does not change the meaning of what I actually asserted, that is, something about the Atlantic coast.

Speech act or "use" theories, in contrast, focus on the actual use of language. Assertions do not usually occur out of the blue, but rather are said by speakers in particular contexts. In a discussion of my recent weekend on the Pacific coast, while reporting something said by a park ranger on a tour of tide pools, I might say "The Atlantic coast has an average temperature of 52 degrees." While this may be what I intended to say (in the context of some Atlantic/Pacific comparison), you, listening to me, might interrupt and ask if

I meant to say "Pacific." You can ask because the meaning of the assertion is tied to the context of its use—but you also recognize the role of intention in my use (did I really mean "Atlantic"?) and you can ask the question in part because the meaning of my assertion points to an ocean we have not been discussing. For Habermas, these three kinds of meaning must all be taken seriously as aspects of understanding language meaning.

When we make an assertion, then, we are making a claim that we take "to follow from" something else (our intentions, some state of affairs, and some situational context). Should someone doubt our claim or not understand it, the nature of the speech act of assertion suggests that a hearer should be able to ask us to explain. If I say that the average water temperature on the Pacific coast is 52 degrees, I imply by the nature of the act of assertion that there are reasons for the claim I make and that under most circumstances I can produce as least some of those reasons. Arguments, the process of generating belief, amount to interactive processes where we make assertions and give reasons for them. Notice that when we give reasons, we generally do so because these reasons are such that, if you accept them, then you must accept the initial assertion as well—that is, the process involves a form of validity. Those who hear us make assertions are similarly disposed: they expect to be able to ask for reasons and that the reasons produced will be valid.

Two conclusions follow from this approach to language meaning. First, argument is necessarily an intersubjective process, that is, an interactive process involving more than one conscious participant or agent. What supports our assertions are reasons generated in the context of communicative action—of bringing someone else to understand and agree or disagree. On this view, the speaker, in making a claim, agrees to provide reasons while the hearer, who recognizes the commitment, is expected to take up a position both receptive to and potentially critical of the claims made by the speaker. Both participants assume obligations in the process and in the outcome of the interaction.

The second conclusion that follows from this approach is that the notion of validity is key to understanding what makes something a reason or warrant for an assertion. For Habermas, "the dimensions of meaning and validity are internally connected" (1998, 228). For each aspect of meaning, in fact, Habermas argues that there is a related aspect of validity. *Intentional validity*, for example, means that an assertion follows from certain beliefs held by the speaker. The meaning of my assertion about the Pacific coast "follows" from what I intend to say—that is, from the beliefs I have about the ocean, etc. When I say "I meant the Pacific coast" in answer to a question about which coast I am talking about, I offer my belief that I had in mind the Pacific as a reason for the assertion I make. In effect, I suppose that you, the hearer, will recognize that if I held certain beliefs, then it must be the case that I meant the Pacific coast.

My assertion also involves a kind of "truth validity," or *formal validity*, in that when I give reasons in support of my assertion, they are given in the following form:

Reason^1

Reason^2

...

Reason^n

Therefore, my assertion.

Implied in this reason giving is the expectation that if the reasons, as premises of my argument, are true, then the conclusion—my assertion—must be true as well. In effect, my assertion implies my recognition of formal validity. Your expectation that my responses will fit this pattern amounts to your acceptance of the same validity framework.

But this same framework also accepts some notion of truth (if the reasons are *true*, we said), which, as it turns out, has the same relational structure as validity. When we accept that some assertion is true based on seeing something, for example, or as a result of some particular experience, it is because there is something (a state of affairs, a reason, a belief) such that if this "something" exists, the assertion in question must somehow follow. In other words, in addition to formal validity, we also expect that there will be a relation of "warrant" between some reasons and something beyond reason—something encountered, something that "is."

A third form of validity is also required by Habermas: "*illocutionary validity*" or the validity of the speech act. From this angle, a speech act—reason or assertion—is valid if it is correctly used given the context of the statement—the who, what, when, where, and to whom of the statement. Unlike intentional or truth validity, illocutionary validity marks the aspect of argument that responds to the actual speaking of language. Like the other forms of validity, however, illocutionary validity also recognizes a certain kind of necessary connection. Given certain social, historical, and environmental factors, some reasons and assertions must follow. While this sort of validity does not require that certain particular assertions follow, illocutionary validity provides necessary constraints about the range of things that can be said. Some assertions will not follow at all and so will amount to nonsense in particular illocutionary contexts. Other assertions will follow, but in so doing will also make questions possible which in turn will give rise to reasons and possible agreement. The illocutionary context of a trial, for example, affords and constrains a range of assertions that can follow. An umpire's call of "Strike!" at certain moments during a baseball game means that the batter has a strike while the same call uttered by a fan does not. Just as a true conclusion must follow from a formally valid argument if the premises are true, the fact of a strike follows in certain valid illocutionary contexts.

Validity, then, in its broad sense, becomes the central logical matter—the relational lynchpin of meaning and the possibility of cooperative action. Habermas concludes: "Validity claims constitute the point of convergence for intersubjective recognition by all those involved" (1998, 232). Validity is the aspect of communicative action that draws people out of their own purposes and concerns and allows them to commit to shared purposes, critique, and creative practice. The result is that older notions of validity (ones that focus exclusively on formal conditions or on "truth" in a narrow sense) can be incorporated into a broader view. In this way, validity is not bound to the formal relations among assertions, but rather is opened up to recognize and make use of other relations between speaker and audience in a shared process of inquiry. "Thus, the conditions of validity no longer remain fixated on the propositional component; room is made for the introduction of further validity claims that are not directed toward conditions of truth (or success), that is, that are not geared toward the relation between language and the objective world" (Habermas, 1998, 232).

The process of making assertions leads directly to the possibility of challenge and the interactive process of argument where those skeptical about an assertion may be brought to shared belief. Shared belief becomes the ground for cooperative action. Inquiry, the process of active transformation of indeterminate situations to determinate ones, can now be seen as an interactive and social process dependent on individual action as well as the possibility of challenge and collective action.

3.6 FALLACIES

While strategic action is not necessarily a bad form of action—it is used to warn people of danger, to give and carry out directions in situations where people have pledged to a larger activity that they may not fully understand—it also carries clear risks. Warnings can become opportunities to motivate people by fear and giving and taking direction can undermine the ability to take responsibility for one's own actions. Strategic action can include coercion and deception that can cause people to act against their own interests or those of others. As we know, the practices of strategic speech acts depend upon the expectation that the primary use of language is for communication, that is, providing opportunities for people to come into agreement and willingly coordinate action. The process of carrying out strategic action against a background of communicative practices gives rise to a class of "argument-like" speech acts that we can call *fallacies*. The linguistic root of "fallacy" is the Latin term, *fallere*, to deceive, and usually names a "deceptive argument" or what the Greeks called a "sophism." Like strategic action in general, fallacies depend upon a background of argumentative practices that are not deceptive in order to carry out their work.

In the history of logic, much attention has been paid to understanding the ways in which arguments can go wrong. Aristotle offered an early classification of argument-like speech acts in *On Sophistical Refutations*. He considers, in particular, the speech acts "entertained by those who argue as competitors and rivals to the death. These are five in number, refutations, fallacy, paradox, solecism, and fifthly to reduce the opponent in the discussion to babbling" (165b10–15).[1] The speeches of the sophists clearly fit the pattern of strategic action in that they aim to force interlocutors into a state of confusion by a variety of means.

More recently, philosophers have sought to provide more concise lists of fallacies in order to aid students in learning how to criticize proposed arguments and to avoid making deceptive ones themselves. Usually fallacies are divided into two categories: *formal fallacies*, which have to do with the structure of an argument, and *informal fallacies* that rely on expectations set by the lifeworld in which the argument is offered. Despite systematic attempts to classify fallacies, there is no exhaustive list. As the logician Augustus De Morgan explained "There is no such thing as a classification of the ways in which men may arrive at error; it is much to be doubted whether there ever can be." To which Horace Joseph, another logician, added, "If we are satisfied that logic should treat of fallacies, it is very difficult to be satisfied with any treatment of them. Truth may have its norms, but error is infinite in its aberrations, and they cannot be digested in any classification" (Robinson, 1947, 188).

Following our discussion of argument as a special form of inquiry in Chapter Two, it is possible to organize fallacies generally around the pattern of inquiry. Some fallacies are related to how a problem is established, some are related to the premises of argument (or the postulates of inquiry), and some are related to the connections between the premises and the conclusion.

SUMMARY OF FALLACIES

Based on the analysis of the process of inquiry, there are three categories of fallacies: ones that relate to *problematization*; ones that relate to the *premises or postulates*; and ones that relate to the *connection* between premises and conclusion.

Fallacies of the Problem

  Fallacies of Exclusion
    False Dilemma
    Excluded Possibilities
    Complex Question

  Fallacy of Mischaracterization

Fallacies of the Premises/Postulates

  Ambiguity
    Equivocation
    Amphiboly
    Division
    Composition
  Petitio Principii

---

[1] All references to Aristotle's work uses the standard paragraph and line numbers from Aristotle, 1941.

Source Fallacies
  Appeal to Authority
  Prejudice
  Laziness
  Straw Argument

Fallacies of Connection

Relevance
  Ad hominem
  Ad baculum
  Ad misericordium
  Ad populum
  Solipsism

Fallacies of the Conclusion
  Argumentum ad ignoratiam
  Post hoc, ergo propter hoc
  Irrelevant Conclusion

Fallacies of Validity
  Fallacies of Intent
  Fallacies of Form
    Denying the Antecedent
    Affirming the Consequent
  Fallacies of Illocution

The first category of fallacies relates to the way indeterminate situations are framed as problems and can be called *fallacies of exclusion*. While setting a problem in response to an indeterminate situation is conditioned by the situation as a whole, some ways of setting the problem cut against a successful solution by narrowing what counts as the problem to be addressed. One way this happens is that the speaker or inquirer describes the problem as a choice between two and only two alternatives. The fallacy is sometimes called a *false dilemma*. The approach takes advantage of the character of choice in that any action selected always excludes some other actions in an absolute sense. In the case of false dilemma, regardless of other alternatives available, the argument presents the situation as a stark choice between two options in a way that ignores or suppresses other viable alternatives. For example, in response to the terrorist attacks of 9/11, President Bush famously presented the problem faced by the world in terms of a dilemma. "We built a vast coalition of nations from all around the world to join us," the President said, "nations which understand that what happened in New York and Washington could happen to them, as well. They understand it's now time to unite to defeat evil. Each nation comes with a different set of capabilities and … a different willingness to help. America says, we don't care how you help, just help. Either you're for us, or you're against us" (Bush, 2001).

It is certainly the case that dilemmas are often accepted starting points for argument. We might agree that a person with a choice between a job offer in New York, for example, and a request to stay at home in Eastern Oregon to care for an aging parent faces a dilemma. The problem with the declaration made in wake of 9/11 was that it was not clear that there were only two alternatives for responding to the problem. Was it the case that there were no neutral or indifferent people in the world? Or was the dilemma really a conclusion to a different problem about whether one can be neutral or indifferent? If the question is the former, then the fallacy is one of exclusion. If the latter, the fallacy is one of mischaracterization (discussed below).

The fallacy of false dilemma is a special form of the general category of the *fallacy of excluded possibilities*. In this wider case, the problem is not set as a dilemma, but is set as a choice among some range of possibilities that exclude (intentionally or not) some other range of alternatives. Suppose a companion of our traveler crossing the meadow agrees that the problem is to cross the ditch, but thinks that it can only be done by finding a shallow ford to cross. If this was the only way to understand the problem, the inquiry would, under the circumstances presented, fail. This sort of fallacy is often associated with problems within established areas of discussion where participants begin by assuming the solution must take a certain form and so block openness to other alternatives. In Peirce's terms, such an inquiry would fail to take advantage of the process of abduction or its most creative form, pure play. Excluding possibilities is a fallacy because it undermines the diverse ways an inquiry or an argument might develop.

A third version of the fallacy of exclusion can be seen in cases where questions are framed in a way that eliminate alternatives by grammatical framing. Someone might ask an official at a prison if the prison policy still allows the torture of prisoners. If the official says "no" then she is also affirming that the prison policy did allow for torture in the past, but no longer does. To say "yes," of course, affirms that the prison policy does allow torture. If the prison never permitted torture, a simple yes or no answer to the question will not do. What is going on is that, in the context of framing the problem, alternatives have been excluded by the form of the question. The fallacy is sometimes called a *complex question*, but it can also be understood as one form of excluding possibilities at the stage of problematizing a situation.

The *fallacy of mischaracterization* occurs when an indeterminate situation is described in a way that sets aside or undervalues aspects of the situation that, if considered, may lead to describing the situation as a different kind of problem. For example, one might describe the firing of an employee as an economic matter and offer, in response, some financial compensation to the former employee for the loss of the job. However, the employee and many of her coworkers believe that the firing was the result of racist attitudes among the supervisors. If understood as a problem of racism, options such as establishing an affirmative action policy or a process of review for dismissing employees might be implemented. These alternative solutions are ignored, not because they are excluded as possible solutions, but because they were never taken as possible in the context of the problem at hand. When a problem is set in the context of a given indeterminate situation, the premises and conclusion are constrained by the problem. When a problem is mischaracterized, the indeterminate situation may not be resolved or it may be resolved in a way that directly produces a new, unsettled situation. It is useful to note that fallacies (like truths) get their character from the lifeworld in which they are offered and the purposes they serve. If a claim is to be communicative action, then unheeded questions (or questions set aside) about how a problem is framed constitute a fallacy—a deception that covers over the alternatives and interferes with the task of understanding.

A second category of fallacies relates to the premises offered in an argument or the postulates that are used in an inquiry. These fallacies can be roughly divided into three subcategories: ambiguity, *petitio principii*, and fallacies of source.

The first fallacy of premises, *ambiguity*, marks premises whose meanings in the stated premise are not clear. In some cases, the terms of a premise are ambiguous because they have no clear referent or because they have different incompatible referents. These are cases of *equivocation*. One might say "everyone to be affected by the proposed tax hike is well off." But the meaning of "well off" is unclear. As a premise for an argument in favor of a tax increase, this seems to be too ambiguous to provide support. In other cases, a term has more than one recognized meaning. Consider the claim "Aliens are dangerous and could destroy civilization." One might reject the claim if "aliens" refers to nonresidents living in the United States and accept it if "aliens" refers to the well-known invaders in *War of the Worlds*. Or the speaker, in an effort to gain assent, might use the premise by suggesting its "Martian" meaning and then use assent to the claim "Aliens are dangerous" to lead to a further conclusion that the U.S./Mexican border should be closed and fortified.

Other problems of ambiguity are grammatical. Consider the claim "A spy photographed a terrorist preparing an attack, but he failed in his mission." In this case "he" could refer to either the spy or the terrorist. An ambiguity of this sort is called an *amphiboly*.

Other related problems of ambiguity occur when parts are taken as wholes and vice versa. One might claim that "each player on the college football team is excellent." When asked to justify this claim, the speaker might respond that "since the football team is excellent, each of its players is excellent." The problem, of course, is that an excellent football team can have players who, by ordinary measures, are not excellent. This sort of mistake is called the *fallacy of division* in which the character of the whole is attributed to the individual parts. This fallacy is related to another, the *fallacy of composition*, in which the character of a part is attributed to the whole. In this case, one might say that she knows many upstanding members of the local Democratic party and so offer as a premise the assertion that the party itself is upstanding. The trouble is that the party may, as a result of the actions of a few members or long-standing policies, be graft-ridden and anything but upstanding. When premises attribute the characteristics of a part to the whole, they obscure the character of the whole.

The fallacy of *petitio principii*, or "begging the question," is the introduction of the conclusion of the argument as a premise. If we begin with the premise, there are no longer cases of racism at our university, it is easy to conclude that there is no racism at the university. This strategy makes the argument easy (since it amounts to the process of showing that some claim, A, follows from A). From the perspective of formal validity, begging the question must be a valid argument since, if the premise, A, is true, then the conclusion, A, must be true; but such an argument fails to give any grounds for the conclusion. If an argument is understood as a process of forming belief or offering an occasion for coming to an agreement, to introduce the conclusion as a premise either says nothing (since all parties already agree) or it contributes nothing to understanding (since it just asserts and reasserts the thing to be shown).

Source fallacies are ones that relate to how a premise is justified. For example, one might offer up a premise and declare that it is justified by an *appeal to authority*. While there may be cases where authorities can justify a claim, there are also cases in which such claims must be called into question. If the president of a major corporation holds a press conference to tell the public that its accounting practices do not allow for falsifying profit figures and then I offer the claim as a premise in an argument about corporate accounting practices, one might challenge my use of the president as a source. Just because he said that the books are not cooked does not mean that they are not. At the same time, should someone report that they have a serious headache and I offer this as a premise in an argument regarding the effects of pollen on some people in the spring, the use of authority would seem to be legitimate. Appeals to authority are not necessarily problematic in an argument, but they suggest that a speaker might be asked for more reasons to justify the use of the premise in question.

Prejudice can also play a role in the selection of premises. In this case, one's interests in the outcome of the argument might lead one to set aside some premises in favor of others on grounds that seem to cut against the goals of communicative action. If I would like to argue against establishing an affirmative action program in my business, I might select a number of supporting premises that affirm that instances of exclusion on the basis of gender and race, for example, are few in businesses like mine. If I do this while setting aside evidence that, despite inclusive hiring policies, most employees from underrepresented groups are in the lowest paying jobs and rarely advance to higher levels in the company, then my selection of premises in light of my prejudices (literally in light of my prejudgment of the issue) undercuts the communicative goals of understanding and agreement. Instead, I am engaged in a strategic action to influence others to reject the policy I oppose. Of course, selection is always a part of the process of argument and selections are controlled in part by my interests and those established in my cultural and societal context. The efforts to avoid source fallacies will consequently involve openness to premises that are relevant to the issue and a general commitment to the goals of communicative rather than strategic action. Prejudices also undermine the process of inquiry in that the resulting selection of resources for transforming a particular indeterminate situation might be inadequate to the task.

A third category of source fallacies might be called *laziness*. Like the other two categories, laziness affects what premises are taken up into an argument or what postulates are used in inquiry. While appeal to authority grounds the selection of premises in some established authority, it does require that the speaker take the time to find out what the authorities hold. While prejudice imposes a highly interested filter on the selection of premises, it also requires that the speaker attend to the premises in question. The fallacy of laziness leads to the selection of premises based on their simple accessibility. Rather than searching for viable premises, one simply takes the first available—the first article found in the library, the first study conducted, the first sources on the first page of an internet search. Like the other source fallacies, the result is a set of premises that fail in both the commitment to communication and the goal of transforming indeterminate situations. If the premises or postulates are those easiest to find, the process of understanding and agreement is hampered in part because information that may make a difference in the process is not actively sought.

A fourth category of source fallacies is the *straw argument* (sometimes called the "straw man"). This fallacy involves using a weak version of an argument in order to criticize it. Like the other source fallacies, constructing a weak argument that is easy to criticize undercuts the goals of communicative action and inquiry. Avoiding this fallacy involves taking seriously the arguments of others and working to find the strongest possible version of the argument under consideration. To take opposed arguments seriously provides for a better understanding of the argument, provides the best possible grounds for fostering understanding and agreement, and uses the best resources for the evaluation of possible courses of action in the process of inquiry. When criticizing arguments that commit this fallacy, it is often best to return to the argument challenged and seek ways in which the critical argument fails to present the original argument adequately.

The final general category of fallacies relates to the *connections* between premises and conclusions. The first of these fallacies are *fallacies of relevance* in which a connection is asserted between premises and conclusion that is not relevant to the inquiry at hand or consistent with a commitment to communicative action. The second subcategory of connection fallacies *relates to the conclusion* and the third form *relates to the three forms of validity* discussed earlier.

The first of the relevance fallacies, *ad hominem fallacies*, argue against a claim by attacking the person making the claim. Suppose that you wish to argue that your employer's accounting practices do not meet Financial Accounting Standards Board (FASB) requirements. To reach this conclusion, you argue that your company president has been known to lie and often fails to report bad news to the stock holders. The argument you have offered does not support your conclusion. Even if the claims about your president are true, they do not address the claim about accounting standards. Instead, you should offer details about ways in which the company's actual accounting practices fail relative to the stated requirements of the FASB.

Another version of the *ad hominem* fallacy is to undermine the claims of a speaker by undermining her or his credibility in a way that it is irrelevant to the claims at hand. To say that a speaker should not be believed in her claims about government fraud on the grounds that she is a woman clearly amounts to an attempt to undermine the premises with an irrelevant (and problematic) challenge to the speaker's credibility. Of course, there are times when a speaker's history and interests clearly matter and at times discrediting a speaker is an appropriate means of criticism. In cases where such criticism supports the goals of inquiry and communicative action, the criticism must be relevant to the premises and conclusion at issue.

A second form of relevance fallacy is called *ad baculum*, that is, the appeal to force. In this case, the connection between premises and conclusion are made not by finding agreement but by forcing the audience to accept the conclusion offered. While this fallacy sometimes takes the form of a direct threat (if you don't accept the conclusion you will be killed or imprisoned), it can also take the form of creating fear in the audience. Though a speaker does not make a direct threat, he might nevertheless suggest that failure to accept the conclusion will have bad consequences such that one should be afraid to dissent. Should a speaker be able to establish a general environment of fear in the context of which dissent is almost always seen as having dangerous consequences, the lifeworld itself becomes one which undermines the possibility of communicative action. In this case, the framing power of the lifeworld literally blocks the process that could transform it.

A third form of relevance fallacy is called *ad misericordiam*, the appeal to pity. When one argues that the conclusion of an argument ought to be accepted because the audience feels sorry for the speaker, the conclusion is not accepted as a result of the relation between the premises and conclusion, but rather because of an irrelevant appeal to a sense of pity. The result undermines the speaker's effort to foster agreement regarding the issues at hand and undermines the ability of the audience to seek understanding and agreement.

A fourth form of relevance fallacy is the *ad populum* fallacy. Here, the connection between premises and conclusion is made by an appeal to "the people." One might, for example, claim that the facts about the incomes and the use of social services of undocumented workers demonstrate that they should be prevented from receiving public support for health care and education. When challenged to make clearer why income and the use of services leads to the conclusion offered, the speaker says that "everyone" knows that they are related in this way. Even if "everyone" does hold such a view, however, it does not follow that this premise leads to the stated conclusion that the workers should not receive public support. Appeals to the people are, on some issues, quite relevant. However, it is often the case that such appeals are not relevant to the argument at hand, and are used to override questions and try to force assent rather than to promote an opportunity for agreement.

A final form of relevance fallacy has to do with a speaker's own role in connecting premises with conclusions. Allan Bloom has identified a form of this fallacy in his book, *The Closing of the American Mind*. The issue, as Bloom observes, is that people (in this case students) often approach arguments as something that is an individual matter and not one located in a context of what we have called communicative action. A person may offer a set of premises and a conclusion and then when asked how the conclusion follows from the premises, she states that the connection is "how I see it," or "my interpretation," where the conclusion is "true for me." While these answers certainly assert that the speaker is convinced of the conclusion, they are not clearly relevant to why the conclusion follows. Suppose a student offers a quotation from Aristotle's *Posterior Analytics* as a premise, "scientific knowledge is not possible through the act of perception" (87b25), and then concludes that, for Aristotle, scientific knowledge is gained through the senses. When someone else observes that the conclusion does not follow from the given premises, the student declares that it does follow because he thinks it does. Should this reason be challenged, the speaker appears to have two options: to give some further reason (which amounts to an admission that the communicative context matters) or to simply reassert the reason already given (which indicates that the speaker does not recognize the context as one of communication). The fallacy—which we can call the *fallacy of solipsism*—appears to accept the goals of communicative action in offering an argument, but then offers the speaker's conviction as the only valid form of connection. Like the use of *ad baculum* and *ad hominem* pseudo-arguments, solipsism sets aside the goals of communicative action. As in these other cases, the best way to prevent such fallacies is to adopt an openness toward challenges and a willingness to seek and give reasons that can be shared in the present situation in the context of a lifeworld.

The second form of connection fallacy relates to the character of the conclusion or outcome of the argument. An *argumentum ad ignoratiam*, or argument from ignorance, identifies as a fallacy the claim that the conclusion follows because it has not been disproven. For example, since there is no evidence that a certain class of people *can* graduate from college, one might conclude, it must follow that no one from that group can graduate from college. This argument fails since it may be the case that no member of the group has ever had the chance to go to college or perhaps every member of the group who has tried college has failed because they lacked the resources to stay through to graduation. This sort of argument is mistaken because it offers only a lack of support for its conclusion rather than evidence or argument in favor of the conclusion.

"*Post hoc, ergo propter hoc*" means "after the fact, therefore because of the fact." In this case, the fallacy is an argument in which one claims that since A precedes B, then A causes B (or "if A then B"). Of course, the fact that B is after the fact of A, does not guarantee that A is the cause. Another version of this fallacy is one in which one finds that A causes B and since A is a member of a larger class, α, and B is a member of β, then α causes β. This is like the fallacy of composition, except that while composition justifies the attribution of the particular to the whole class, the *post hoc* fallacy recognizes the attempt to justify the class that follows by attributing its cause to the class that is taken as coming first. In either the simple case or the complex one, the *post hoc* fallacy treats a sequential relation as if it is causal and so risks overlooking other aspects of the situation that are causally involved.

Finally, it is a fallacy of the conclusion when an *irrelevant conclusion* is asserted as a response to the situation at hand. In begging the question, the argument simply reasserts the premise or premises as the conclusion and so fails to add anything that will transform the original situation. In the case of the fallacy of irrelevant conclusion, regardless of what premises are offered, the conclusion asserted is irrelevant to the situation and so likewise fails to transform the original situation.

The final category of fallacies relates to the three forms of validity: intentional, formal, and illocutionary. An intentional fallacy is to take premises and conclusions in ways not intended by those who have offered them. To take a misstatement, for example, as what the speaker actually meant when it is clear from context or questioning that the speaker meant something else, is to commit a *fallacy of intent*. Such a fallacy cuts against the possibility of communication in that it ignores or obscures the meaning intended by a participant in the conversation.

Formal fallacies are ones that relate to the structure of arguments. Consider again the argument form *modus ponens*. Suppose that if Mary mows the lawn, then she will be late for dinner and she does, in fact, mow the lawn. It follows that if both claims hold then Mary must be late for dinner. Now, suppose that the claim "if Mary mows the lawn, then she will be late for dinner," holds and we find out that she did not mow the lawn. One might be tempted to conclude that Mary therefore was not late for dinner. The conclusion, however, does not follow necessarily. Mary may not have mowed the lawn, but instead may have gone to the coffee shop down the street, got distracted, and was late for dinner. Arguments of the form

$$s \rightarrow p$$
$$\neg s$$
$$\therefore \neg p$$

are fallacies of *denying the antecedent.*

In a similar way, if it is true that if Mary mows the lawn, then she will be late for dinner, we might find that Mary was, in fact, late for dinner. In this case, we might conclude that Mary must have mowed the lawn. However, it could be the case that Mary was late for dinner for some reason other than her mowing the lawn; she may have gone to the coffee shop and stayed too long. Arguments of the form

$$s \rightarrow p$$

$$p$$

$$\therefore s$$

are fallacies of *affirming the consequent*. In either denying the antecedent or affirming the consequent, the structure of the argument is such that the premises do not connect with the conclusion in a way that guarantees that if the premises are true then the conclusion must be true.

A final class of validity fallacies regards the relationship between the claims made and the context of the speaker. In general, when a justice of the peace pronounces a couple married in the right circumstances (with the proper documents signed and witnesses present), then a marriage is established. However, if the same justice of the peace reads the lines in a community theatre play or says the words while sitting with friends playing cards, no marriage is established. Or imagine that at a protest against layoffs, a person steps to the microphone and shouts that unless the jobs are restored the remaining workers will immediately strike. If the speaker is a labor representative who has come to the microphone following a meeting of the union in which a strike vote was taken, then the claim follows as a valid illocution. If the speaker is a member of the union but not a designated leader, then the claim might amount to a hope or a threat, but it will not be a valid illocution—the speaker is not properly related to the claim she makes. Cases where speakers make assertions as if they are eligible to do so given the circumstances but are not are *fallacies of illocution*.

In all of these cases, the named fallacies mark rough categories that can be used to examine arguments and decide whether or not they satisfy the pattern of inquiry on one hand and the commitment to communicative action on the other. At times, practices that are ruled out because they are fallacies are not clear cut in their application. Rigidly holding that certain concerns are fallacious at all times is also a fallacy in that it undercuts the opportunities for speaker and audience to explore the issues that enable and constrain premises and conclusions. Sometimes it does matter who speaks and who listens. Sometimes appeals to the people or to pity or fear are appropriate ways of understanding a problematic situation and in framing how we can respond in inquiry and argument. In the end, perhaps all of the fallacies can be understood as violations of the general principle proposed by Peirce: "Do not block the way of inquiry" (1.135).

3.7 EXERCISES

Consider each of the following examples of fallacies. Decide which fallacy is committed in the example. Explain your conclusion for each.

1. The mayor of a small Oregon town declared that her proposal for land use policy was the finest in the state. When asked for reasons, she explained that the chair of

the state land use board said that it was the finest. It turned out that she was also the chair of the state land use board.

2. The police arrested the burglar for stealing computers from residence halls on campus using a stolen key. The burglar demanded to be released on false charges since the key he used was not stolen but made from an impression of the locks.

3. When faced with the prospect of intervening in another nation's troubles, we can either take forceful action or do nothing. If we do nothing, the troubles could spread. So we must take forceful action.

4. In a recent philosophy course a student argued that substance must be understood as one and not many. The reason for this, he explained, was that anything that was many could not be one.

5. The salmon runs will come back naturally over time. I tagged several fish three years ago and one returned this spring. Since one salmon survived, we can be confident that the salmon run will survive.

6. We have considered the nature of being from all angles. We seem to have three possibilities. Being could be one thing or many or it could be many ideas unified in one divine conspectus. In order to avoid the traditional problem of the one and many and have a workable theory of being, the third option is the only viable alternative.

7. J. S. Mill wrote: "[Each] person's happiness is a good to that person, and the general happiness, therefore, [is] a good to the aggregate of all persons."

8. A student was assigned the task of researching Charlotte Perkins Gilman. "Gilman," the report began, "was a religious figure known as the goddess of the home." Puzzled by the report, the teacher grading the assignment entered Gilman's name in a web search. The first entry was titled "Domestic Goddess."

9. It has been argued that a new government policy to grant amnesty to undocumented workers in the U.S. will solve some of the problems related to the cost of social services. A recent survey shows that 53% of American citizens and permanent residents do not believe that amnesty will solve any problems. This lack of support demonstrates that the amnesty plan will not solve any problems.

10. Lewis Carroll wrote this short dialogue: "Who did you pass on the road?" the King went on, holding is hand out to the messenger for some hay. "Nobody," said the messenger. "Quite right," said the King, "[Alice] saw him too. So of course Nobody walks slower than you."

11. A liberal congresswoman spoke on the floor of Congress about the fate of health care. "The conservatives," she said, "are interested in killing the children of the poor. Look at the facts. They have cut funding to children's health care each year." A conservative responded: "The bill we proposed aimed at reducing the deficit so that the economy would improve and give better opportunities to the poor. You are wrong about our motives." "My accusation," the liberal replied, "is true because it squares with the facts as I see them."

12. Descartes' argument for the existence of God is basically that if I think that there is a God, there must be one. This is clearly a mistake since, when I think of a million dollars in my bank account, it does not mean that I have a million dollars.

13. The owner of a convenience store noticed that a large number of candy bars were missing after the night shifts on Tuesdays, Thursdays, and Saturdays. On investigation, the owner discovered that the same cashier was working each night. The owner thought that the cashier was a disreputable looking character, but she was the only employee willing to work the late shift. While reviewing the security tapes, the owner also noticed that the same seven people came in each night and passed slowly by the candy display. The owner concluded that the cashier was stealing candy and fired her.

14. If cell phones lead to reckless driving then we would see an increase in accidents. And we have seen an increase in accidents. Therefore cell phones lead to reckless driving.

15. Ella listened to the presentation by the real estate developer. When it was her turn to speak, she said that she had known the developer for many years. His mother brought dinner to her family once when Ella's father was ill. She said that the real estate agent was a veteran. Finally, Ella concluded that the developer's proposal to build new affordable housing on the site of an old elementary school should be approved.

16. The reason I haven't called or returned your calls is because I thought we were no longer together. Your sister told me that you were sick of this relationship and that you wanted to break-up or at least take some time apart. So for my part, I was only doing what you wanted.

17. The press secretary announced that it was likely that Uzbekistan had nuclear weapons capability. A reporter asked for a reason for his claim. The press secretary explained that he had been up for over 24 hours and was exhausted and upset about the situation. A second reporter asked again for a reason. The press secretary shook his head and said that if we failed to understand the risk now, we could all be destroyed.

18. If you support the conservative cause you need to vote for the Republican candidate in the next presidential election. It is true that the Republican candidates will be far from being ideal conservatives, but they will be more conservative than the Democratic candidates and therefore they will be all we have to choose from.

19. Going ahead on the new reality show is a moral issue. Even if it displays graphic violence that may offend many people, we must go forward because we have a moral obligation to our shareholders to turn a profit. Our ratings are terrible and this show will change all that. We are morally obligated to air the program.

20. Regardless of how much higher the ratings are when we air the new show as opposed to when we don't, shamelessly displaying scantly dressed women for no real reason is disrespectful and offensive to our fans. We need to think about our fans first because they are the ones whose viewership supports this station and if they start tuning out we'll be out of business very soon.

21. This movie is for sure going to be great. First of all David Peoples wrote it and he is a great screenwriter. Secondly, just about every leading and supporting actor in it has won Academy Awards. And lastly, I loved every movie that the director and producers of this movie have made.

22. Officer Johnson pulled me over for speeding, but, your Honor, I give you my word that I was not. My word on this is good; just ask anyone who knows me. Officer Johnson, on the other hand, when you ask people who know him well, say that this is a guy who is on his third marriage, is disliked by all his neighbors, and is a flat-out horrible human being.

23. Higher paying managerial positions should be given to those who deserve them the most. Sean is very poor, his family has gone through many unspeakable tragedies these past few years, and recently he looks like he is starving. If anyone deserves a high paying manager position it's him.

24. The Universe could have been created in only one of two ways; either randomly and out of nothing (i.e., what some so-called scientists refer to as the Big Bang) or by a loving and caring God who continues to look after us.

25. It is important for all shop owners to protect their shop. Most shops in this neighborhood get vandalized daily, unless the shop owners pay thugs for "protection." It may not be good idea to pay thugs for protection, but let me state this as clearly as I can to all you shop owners around here: paying thugs is a whole lot better than the alternative.

26. If Eloise did not study diligently, then she would fail the course. I heard that she failed, so she must not have studied diligently.

27. It is never good to talk to pitchers in between innings when they are pitching well, especially when they are throwing a no-hitter. I've seen many near no-hitters, but never a no-hitter. Every time I've seen a pitcher come close, some fool always comes and says something to him in between innings and just like clockwork the pitcher gives up a hit in the following inning.

28. The memo from the manager said that we should close early if a severe storm is expected. Though the weather report predicts good weather all week, I expect bad weather. So following the manager's directive, we will close early.

29. If you take care of yourself when you are young, then you will live a long life. But you did not take care of yourself when you were young. Therefore you will not live a long life.

30. Chuck Norris is the best actor ever and a master of the martial arts. The show he starred in, *Walker Texas Ranger*, has taught me many invaluable moral lessons and it's a great show too. So when Chuck says we need to oppose amnesty for any and all illegal immigrants, I'm with him and you need to be too.

31. Are you seriously considering voting for any one of the candidates on the Democrat slate when you know they are just going to raise taxes and run our wonderful state economy into financial ruin?

32. In Cervantes' novel, Don Quixote came across a plain dotted with windmills. "What luck," he declared to his companion, "we are about to encounter thirty giants that I will fight and kill!" His companion, Sancho, objected that these were not giants but windmills. Don Quixote replied that his companion knew nothing of adventures. "These," he insisted, "are giants and I shall fight them." He attacked the first and was unhorsed by the spinning sails of the windmill.

33. I spoke with Madeline. She was leaving her office to get her car from the parking ramp. She said that she was afraid to get it since it was late and the ramp was not well lit. I said I wouldn't be afraid and she shouldn't be either.

34. I have never seen something alive come from something not alive. Evolution depends upon this sort of thing happening. But since it can't happen, we can reject evolution as a viable theory.

35. Children throughout this country, especially in poorer areas such as the inner cities and rural areas, are falling farther and farther behind. Therefore, we need to support the "No Child Left Behind" program and its rigorous testing program.

36. If Malcolm was born in the U.S., then he is a citizen. He is a citizen, so he must have been born in the U.S.

CHAPTER 4

THEORY OF THE SYLLOGISM

4.1   NOMINALISM, REALISM, AND ABDUCTION

How are we to understand the "grounds" that provide warrant for the reasons we give in argument? Are these grounds strictly the result of convention or social history or particular choice? Or are they concrete givens that are what they are regardless of human social development or choice? Or are they some combination of these things? As we saw in the last chapter, Habermas identifies a shared "lifeworld" as context that warrants our reasons. Speakers, Habermas concludes, "acting communicatively always come to an understanding in the horizon of a lifeworld. Their lifeworld is formed from more or less diffuse, always unproblematic, background convictions. This lifeworld background serves as a source of situation definitions that are presupposed by participants as unproblematic. In their interpretive accomplishments the members of a communicative community demarcate the one objective world and their intersubjectively shared social world from subjective worlds of individuals and (other) collectives. The world-concepts and corresponding validity claims provide the formal scaffolding with which those acting communicatively order problematic contexts of situations, that is, those requiring agreement, in their lifeworld" (1984, 70). Does this mean, however, that the "unproblematic" aspects of a lifeworld are solely products of social evolution and human activity? Even as validity is bound up with the use of language, it is also bound up with fundamental questions of ontology—of what there is. Historically, this question, especially as it applies to argument, was understood as the debate between nominalism and realism.

Nominalism is the view that things grouped together under the same name share only that—the name.[1] Groups or classes understood nominally are assembled (or selected) by the actions of those who assign the names and do not have any "underlying" or "real" commonality apart from the fact that they have been given a common name. There are, on this account, no "natural" kinds, but rather only "kinds" that are a product of some naming process. Nominalism, like Habermas's "agreed-upon interpretative patterns, loyalties, and proficiencies" (1998, 237) usually credits human action as responsible for the makeup of named groups such as the groups human beings and cats. As a result, nominalism is often viewed as a form of human social construction. *Realism*, in contrast, holds that kinds or classes are real in the sense that they exist prior to any naming process so that the kinds, *human being* or *triangle*, for example, exist independently of any process of selection. This does not, of course, mean that specific individuals necessarily exist, but rather that their kinds do and, as a result, are what makes it possible for particular things to be human beings or triangles.

The concept of abduction as developed by Peirce is part of a response to the long-standing philosophical conflict between advocates of "nominalism" and "realism." Peirce strongly rejected pure versions of both of these views and, in light of the process of abduction, argued instead that kinds are the product of an interaction between an active agent or knower and the things and events known. On this account, the things that agents encounter suggest connections and separations. Green plants, for example, suggest connections to each other through their common process of photosynthesis while various fish suggest connections to each other through their characteristic environments (in water) and so on. Abduction recognizes that, in at least some sense, characteristics related to the process of classification are brought to the interaction by the things and events classified and so abduction is consistent with a form of realism. At the same time, the agent or knower also exercises a process of selection and division in interactions so that, for example, human visual capabilities recognize a certain range of colors and human interests that often lead to the selection of some characteristics over others. As a result, plants, for instance, are classified in a particular way based in part on their physiology and environments and in part on their ability to reflect light in a certain way. At the same time, they are classified by the ways in which reflected light is received by human visual perception and by the particular interests of knowers that might include such things as the edibility or toxicity of the plants.

---

[1] While the traditional name of this view is "nominalism," in recent debates about realism, the position contrary to realism has often been called "anti-realism," leaving open just what the alternative might claim. Nominalism—as suggested by its name—affirms a more or less specific view that the alternative to realism involves a process where things get the character of being trees or people through selection and naming at the discretion of the ones naming.

The process of abduction, Peirce argued, works because human beings are continuous with the world that they attempt to know. Since human beings are affected by physical processes such as gravity, their experiences suggest relevant hypotheses about how to understand physical laws and these hypotheses will tend to be more reliable than random guesses. This same commonality of environment and embodiment provides humans with a starting point for thinking about other kinds of beings who share the same environment. As a result, things and events bring something to the interactions involved in inquiry. At the same time, human "affordances" (our visual and aural apparatus, the structure of our brains, our physical size and motility, for example), and interests constrain the process. The continuity between human agents and other things and events make abduction possible and, to some extent, dependable.

When human beings use abduction, they begin to generate classes or kinds that help them to negotiate the world. These groupings are developed in the context of particular interactions within lifeworlds, since much of what we learn about classification is not through abduction directly but by other processes that are tied to language use and social interactions with other human beings. Of course, even the things we learn from others begin in abduction at some point and, as a result, knowledge literally emerges from on going interactions among people, and between people, non-human others, technologies, ecosystems, and so on.

The implication of all of this for logic is that abduction—in its particular settlement of the nominalism/realism conflict—generates the kinds and classes with which we inquire. From this perspective, inquiry begins with abduction and sets the stage for the processes of reasoning and testing. By suggesting ways of sorting or ordering a situation, abduction frames how we think about a given situation and so also how we test possible efforts to transform the situation. Abduction, in short, provides us with more or less spontaneous classifications and orderings of things and events that we can think about, enabling us to move from indeterminate situations to determinate ones. In this process, ordering and classifying becomes expressed in *assertions*. Assertions, on the view offered here, refer in their use to the order of relations one finds in experience. Assertions in relation to other assertions provide a kind of map of areas of the world. As with maps, assertions (again, in their use) carry with them a network of relations such that when we have a particular claim, we have or potentially have a whole network of relations that can direct further action both in the form of reasoning and in the form of overt action.

To say, for example, "All human beings are animals" is to express the inclusion of human beings within a class that is at least as big as the class of human beings, the class of all animals. At the same time, the assertion also serves as a guiding principle or a literal map with which to direct action. To operate with the assertion is to be disposed to act toward human beings and animals in some common way. Put another way, assertions have two functions.

The first function of assertions, *denotation*, establishes expectations for further experience. For example, "All human beings are animals" establishes the expectation that when human beings are encountered, the encounter will involve certain modes of action. To claim that all human beings are animals leads one to expect that human beings will have certain needs and limitations comparable to those of other non-human animals and that interaction will involve some particular kinds of action. The second function of assertions, *connotation*, suggests that when one operates with the assertion "All human beings are animals," a series of other assertions follow such that these further assertions provide reasons for conclusions that go beyond the initial assertion and its disposition to act. In the context of a process of inquiry, connotation provides the resources for proposing hypotheses and testing their consequences in thought. Denotation guides the transformation of reflective action (thinking about things) into outward action that can experiment with hypotheses and transform indeterminate situations. If assertions have this "double-barreled" meaning (to borrow a phrase from William James), then the subject matter of logic when it is focused on *reasoning* and *testing* is the combined denotative and connotative functions of assertions. The combination of the denotation and connotation of a given assertion will be called a *proposition*.[2]

There is a long history of philosophical debate over the nature of propositions. One common way to describe propositions is given by Bertrand Russell: "A proposition is something which may be said in any language: 'Socrates is mortal' and '*Socrate est mortel*' express the same proposition. In a given language it may be said in various ways: the difference between 'Caesar was killed on the Ides of March' and 'it was on the Ides of March that Caesar was killed' is merely rhetorical. It is thus possible for two forms of words to 'have the same meaning'" (1962, 9–10). Therefore, the proposition represented by the assertion, "Snow is white," is identical with the proposition represented by "*La nieva es blanca*." The problem for such a view is the need to identify "where" such propositions exist. If they exist in the mind or brain, then there is a problem of how propositions can be part of the world outside the mind or brain and how people can have access to each others' propositions. If propositions exist in some realm outside human beings, then they seem to generate metaphysical problems about where this realm of propositions is and how anyone can have access to it.

The view I adopt here is that propositions are understood functionally in terms of habits that frame thought and action. The connotation of the assertion "The snow is white" is found in the assertions that it calls up when it is active in consciousness (i.e., when we think about it). These "patterns of activation" in our thinking are habits. Similarly, the denotation is found in the

---

[2] Assertions are speech acts (written or spoken) that make a claim. The *meaning* of speech acts, then, have two aspects: *denotation* marking the connection between assertions and further outward action and *connotation* marking the connection between assertions and other assertions.

various patterns of behavior that are carried out in the encounters with the things described. When we encounter certain things in experience, we have complex physical responses that are ordered by habits. Some of these habits also call up assertions that provide resources to direct or channel the flow of our behavior beyond the present physical responses. From this perspective, there are no problems "reconnecting" propositions to the world since they are concrete habitual practices already in the world. There is also no problem of accounting for connections among metaphysical "realms" because propositions are processes of interaction that support reflection and direct behavior. Other people (and the rest of the "external" world) can at least partially share our propositions in that they encounter connotation through our arguments and denotation through our actions and the ways in which we transform situations.[3]

In the discussion so far, we have talked about assertions referring to groups of things and events where the collection is a product of both what the things and events "have" independently of the interaction and what the agent or knower brings to the interaction. A more traditional understanding of assertions is that they refer not to groups but to *properties* that "inhere" in a thing. On this view, "All human beings are animals" is actually a claim about some properties that are in things that are human beings—in this case, the property of "animalness." Put another way, all the beings that have the property of being human also have the property of being animals. A particular thing, say Hillary Clinton, is something that has the property of being human and also the property of being an animal. This character of Clinton is not a matter of her being classified as a human and animal, but rather that she has the real properties of each, which we recognize and assert. This approach can be called *predication* where groups are formed "naturally" by things that share common *properties*. This way of understanding groups is closely related to the realist position we discussed earlier where there are "real" aspects of things that make them people or animals and that these things exist prior to any sorting an inquirer might carry out. An alternative approach, *classification,* is closely related to the nominalist position in that it seems that things can be put into whatever *collections* or classes we like and, as a result of that sorting, stand in relation to other things.

The approach I take (though I most often use the language of classification), informed by Peirce's response to realism and nominalism, combines both of these. The notion of abduction recognizes predication as capturing the reason that abductions are not merely wild guesses but are suggestions that emerge from the interactions at hand. Group membership is prompted by

---

[3] If propositions are the denotation and connotation of an assertion, they will be partially constrained by the language of the assertion and this aspect may or may not translate easily into other languages. Even the simple assertion, "Snow is white," may have connotations in English that one does not find in a roughly comparable claim in Aleut. What makes the claims "roughly" comparable, however, is the denotation of the assertion which is manifested in action. Hence propositions can at once mark commonalities across lifeworlds and also divergence among them.

things and events and, in this sense, is a result of the "properties" that inhere in the things sorted, that is, by what is encountered. At the same time, abduction recognizes that properties do not fully determine class membership because the operation of classifying selects as well (constrained by resources, interests and so on). On this view, assertions mark both properties and selection.

Royce, in his theory of order as presented in *The Principles of Logic*, argued for a comparable approach to the relation between realism and nominalism. "What constitutes order, and what makes orderly method possible, is not the product of the thinker's personal and private caprice. Nor can he [or she] 'by taking thought' willfully alter the most essential facts and relations upon which his methods depend. If an orderly classification of a general class of objects is possible, then, however subjective the choice of one's principles of classification may be, there is something about the general nature of any such order and system of genera and of species—something which is the same for all thinkers, and which outlasts private caprices and changing selections of objects and modes of classification" (1914, 316).

The validity of assertions inferred from situations and the validity of arguments in light of this idea of meaning then depends in part on the ways in which what there is leads to the assertions made. The valid connection between a state of affairs and an assertion—the truth of the assertion—depends upon the circumstances of interaction that give rise to the claim. That Hillary Clinton is a human being follows from both what Clinton brings to her interactions and what is selected by those who encounter her. But what of the validity of argument? What constrains the connections between reasons and assertions such that some structures are valid and others are not?

One might conclude that formal validity is the result of agreement where we simply have come to agree that the following argument is valid.

> All humans are animals.
>
> All animals are mortal things.
>
> So, all humans are mortal things.

Yet it seems that the relations captured here are more than simply a matter of agreement. Arguably, what makes validity "the point of convergence for intersubjective recognition by all those involved" is not its "agreed-upon" quality but a kind of structural necessity that owes to something beyond reason. From the standpoint of abduction, we might say that while the particular form of the argument given is conventional, the relation it expresses seems to emerge from the interactions themselves. Nature, Peirce might say, suggests the valid structure not simply as a convenience to inquirers but because the structure of things implies it.[4]

---

[4] See Peirce, 2.713.

4.2   THE THEORY OF THE SYLLOGISM

Much of what logicians say about validity is a product of long discussion that began in Western culture around the work of Aristotle whose systematic reflections on argument provide a starting point for understanding the structures of validity. For Aristotle, arguments are, at their simplest, to be understood as syllogisms. "A syllogism," he says, "is discourse in which, certain things being stated, something other than what is stated follows of necessity from their being so" (24b15). The basic form of a syllogism is one which includes two premises "affirming or denying one thing of another" and a conclusion—some further statement of affirmation or denial not included among the premises but which "follows of necessity."

Aristotle discusses his theories of categories, propositions, and the syllogism in the first five books of the collection of his works that have come to be called the *Organon: Categoriae, De Interpretatione, Analytica Priora, Analytica Posteriora,* and *Topics.* Aristotle's works probably began as lecture notes rather than treatises, perhaps actually written down by his students. Despite this apparent remove from Aristotle's own authorship, the texts of the *Organon* have long been taken as "authentic" and serve as the basis both for our understanding of Aristotle and the origin of the study of logic in Western philosophy. After his death in 322 BCE, Aristotle's works were acquired by a young follower, Theophrastus, whose student, Neleus, inherited them. When Neleus died, his family tried to preserve the works by storing them in a pit where they were damaged by moisture. Eventually (around 100 BCE) they were rediscovered and (in 86 BCE) fell into the possession of the Roman general, Sulla, and then to a series of scholars who tried to restore and preserve the works. Eventually, in the 9th century, Muslim scholars acquired copies and wrote a series of commentaries that helped to sustain interest in Aristotle's work. In the 13th century, largely through the work of Thomas Aquinas (depending in significant ways on the Islamic commentaries), Aristotle's works were incorporated into Christian thought and made increasingly available. The *Organon* was the definitive starting point for Medieval logicians and Aristotle remained the central logical theorist for Western philosophy until the 18th century. His work persisted in its influence in part because it presents a view of logic and its relation to the world that people have found useful and persuasive.

By the 20th century, Aristotle's logic tended to be interpreted in three ways. In the first, based on an understanding of the theory of the syllogism, it has been taken as a purely formal theory of how human beings think. From this angle, the formal relations among propositions are expressions of the operations of mind alone. This view (though it persists in the study of logic in some quarters) has largely been supplanted by the second view, that the theory of the syllogism is only a subset of first-order quantificational logic. The argument is that Aristotle did not have the intellectual resources to develop a theory of a more general sort that advances in logic during the 20th century made possible. As a result, the theory of the syllogism is set aside as a simpler and less complete version of logic. The third approach to Aristotle's logic, still very much in evidence, is to see syllogisms as key components in understanding the operation of successful argumentation. From this angle, Aristotle's theory gives us a picture of successful arguments and, as a result, a standard in terms of which to criticize attempted arguments as well. Aristotle's work, however, suggests that none of these approaches are very accurate descriptions of his own view.

For Aristotle, logic was not strictly about thought or mere form abstractions or even the practices of successful argument, but rather drew together ontology, epistemology, and formal concerns. That is, the orders that emerged in the study of propositions and syllogisms provided one approach to the order of things and events. To know about particular things and events is always to understand them in relation to classes. Things and events are not merely what is said but—as I have argued—something emergent in interaction influenced by what is brought to the interaction by the things and events involved and by the process of selection brought by those caught up in the interaction. The general structure of relations captured in the forms of assertions ordered in syllogisms are at once a picture of the ways in which things in the world can stand in relation to one another and the way in which things are known. To investigate the world is, in a real sense, to investigate the structures that are found in thought and to investigate the structures of thought is to investigate the structures of the world as they emerge in the processes of interaction. In the following sections we will consider the details of the theory of the syllogism, keeping in mind that the way I understand assertions and their relations is framed by the ontological perspective offered above.

4.3 STANDARD FORM PROPOSITIONS

Aristotle begins his work, *Topics*, by making a distinction between argument as demonstration and argument as dialectic. "[An argument] is a 'demonstration' when the premises from which the reasoning starts are true and primary, or are such that our knowledge of them has originally come through premises which are primary and true: ... [argument], on the other hand, is 'dialectical', if it reasons from opinions that are generally accepted" (100a25–30). Aristotle explains "Things are 'true' and 'primary' which are believed on the strength not of anything else but of themselves: for in regard to the first principles of science it is improper to ask any further for the why and wherefore of them; each of the first principles should command belief in and by itself. On the other hand, those opinions are 'generally accepted' which are accepted by every one or by a majority..." (100b18). Although much is said later in the history of logic about this distinction, the distinction itself is closely related to the concerns raised by Habermas.

In demonstration, we find a response to questions that demand support for our assertions. If we are to claim that all human beings are mortal, we seek reasons that are valid in Habermas's sense. In dialectic, we see both the process of testing assertions by framing them carefully and examining what follows and we see as well the possibility of new ideas emerging as old ones are examined and challenged as part of the cost of communicative action. Yet, for Aristotle, what frames both forms of argument is the structure of valid argument itself—the syllogism. The difference in interest, he says, "will make no difference to the production of a syllogism in either case; for both the demonstrator and the dialectician argue syllogistically after stating that something does or does not belong to something else" (24a25). In the following discussion, we will look at the structure of validity as it develops in the theory of syllogistic argument.

Arguments, of course, are formed by a sequence of assertions, some serving as premises and at least one standing as a conclusion that is to follow if the premises are true. An *assertion*, understood theoretically in the sense discussed above, is composed of two terms, the *subject* and the *predicate*. The terms are joined by a *copula*. Consider

All human beings are animals.

Here, 'human beings' names the subject term, 'animals' names the predicate term, and 'are' names the copula (always a form of the verb 'to be'). The *standard forms* of assertions also include

Some human beings are women.
No human beings are fish.
Some human beings are not men.

In each case, the assertion consists of a subject (human beings) and a predicate (women, fish, and men) and an appropriate copula. Each assertion also includes words to indicate *quantity* (e.g., 'all' and 'some') and a term to indicate whether a term is *denied or affirmed* (i.e., 'no' and 'not' or their absence).

In the case of assertions in nonstandard form, it is often possible to rewrite them in standard form by keeping track of the subject and predicate expressed. Consider:

Bicycles always have two wheels.

In this case, "bicycles" is the subject term, but it is ambiguous since it could mean some bicycles or all bicycles. To rewrite the statement in standard form, we will specify that the assertion applies to all bicycles. There is also no copula given so it is common in English to rewrite the predicate so that a copula can be added. In this case, the predicate is "two-wheeled things." When combined with a copula, the proposition can be expressed

All bicycles are two-wheeled things.

Consider

Among the cats are all tigers.

In this case, the subject term follows the predicate term so that the standard form of the assertion is

All tigers are cats.

Each assertion in standard form has three further characteristics that affect its meaning. These are *quantity*, *quality*, and *distribution*. Standard form assertions have one of two quantities (as suggested above) marked by the words 'some' and 'all'. Assertions of the first sort are called *particular* and of the second sort are called *universal*. Although "some" may mean *at least* one thing (as in "Some chickens lay eggs"), we will take it to mean "at least one and perhaps all." Universal assertions, those marked by 'all', on the other hand, refer to every one of the things that have the property in question (or, alternatively, everything in the designated class). Consider the following assertions:

(1) All daisies are flowers.

(2) Some mammals lay eggs.

(3) A few bicycles have four wheels.

(4) Every tennis ball is spherical.

Statement (1) is universal; (2) is particular. Statement three, in standard form, is "Some bicycles are four wheeled" (e.g., bikes with training wheels) and so is particular. Statement (4) is rewritten "All tennis balls are spherical" and so is universal.

Consider

Barack Obama is a human being.

One might think that the statement represents a particular assertion since it only involves one person, Barack Obama. In Aristotle's analysis, however, *this is a universal assertion* in that the subject includes all of the class of things named Barack Obama (or all the things with the unique set of properties that make up Obama). In general, any assertion whose expression involves a subject designated by a proper name is a universal assertion.

The *quality* of an assertion can be defined as either affirmative or negative. If it is *affirmative*, it *affirms* something of the subject. If it is *negative*, something is *denied* of the subject. Consider

(5) All citizens are loyalists.

(6) Some citizens are taxpayers.

Statement (5) is a universal affirmative statement because it affirms of the entire class of citizens that they are loyalists. (6) is a particular affirmative in that it affirms of some citizens that they are taxpayers.

(7) No citizens are foreigners.

(8) Some citizens are not lawyers.

What of (7)? In this case, the statement is universal in that its assertion applies to every member of the class of citizens. Since it also denies that citizens are foreigners, its quantity and quality are universal and negative. (8) applies to some members of the class of citizens and so is particular and, regarding those citizens, it denies that they are lawyers. This is a particular negative.

We can refer to these four standard forms of assertions as follows:

Let universal affirmative assertions be labeled 'A'

Let universal negative assertions be labeled 'E'

Let particular affirmative assertions be labeled 'I'

Let particular negative assertions be labeled 'O'[5]

Finally, the terms of each assertion also have the character of being distributed or undistributed. A term is *distributed* if it attributes something to *all the things or events in the group* (kind or class) to which it refers. A term is *undistributed* if *its attribution to a group is indefinite*. In the cases of all A and E assertions, the subject term is always distributed since it refers to all members of the listed class. For I and O assertions, the subject is undistributed because it applies to an indefinite group of the listed class.

What about the predicates of affirmative assertions? Consider "All human beings are animals." While the assertion clearly attributes something to all human beings (the subject), the reference of the predicate is to some (perhaps all) of the listed class, in this case the class of animals. As a result, the predicates of A assertions are undistributed. The predicates of I assertions are also undistributed. For example, "Some citizens are taxpayers" applies to some indefinite group of citizens (the subject) and some indefinite group of taxpayers (i.e., not necessarily all in either case).

Consider negative assertions. In E assertions such as "No citizens are foreigners," the subject is distributed. Notice in this case, however, the predicate is also distributed in that it applies to every one of the predicate class. Take the class of foreigners: if every member of the class is considered, not one is a citizen. The predicate is distributed over the entire class. In the case of O assertions, the subject is undistributed (because it is only some of the class), but the predicate, again, is distributed. The predicate of an O assertion attributes something to every one of the members of the named group. So "Some airplanes are not two-winged" says that, of some indefinite group of airplanes, they are not among *any* of the two-winged things. If we were to set out every two-winged thing, we would find no monoplanes among them. The predicate is therefore distributed. Figure 4.1 summarizes the distribution of the subject and predicate terms of standard form assertions.

It is useful to suggest a notion of truth that can be used to make sense of the ways in which the premises of a syllogism can be true. Following Habermas and Dewey, we can say that *an assertion is true when it is warranted*, that is, it follows appropriately from intentions, states of affairs or reasons,

---

[5] Cohen and Nagel observe that "The letters *A* and *I* have been used traditionally for affirmative [assertions]: they are the first two vowels in *affirmo*; while *E* and *O* symbolize negative [assertions]: they are the vowels in *nego*" (Cohen and Nagel, 1934, 37).

Standard form	Subject	Predicate
A "All corporations are taxpayers."	Distributed	Undistributed
E "No corporations are taxpayers."	Distributed	Distributed
I "Some corporations are taxpayers."	Undistributed	Undistributed
O "Some corporations are not taxpayers."	Undistributed	Distributed

Figure 4.1   Summary of Distribution for Standard Form Assertions

that is, when it is a valid assertion in Habermas' sense. That an assertion is true in this way is always open to challenge and new inquiry. *An assertion is false when there is warrant for its negation.* If, for example, one validly asserts

(10) Some Democrats are in favor of the Iraq War.

then there is warrant for the assertion

(11) It is not the case that no Democrat is in favor of the Iraq War.

where "It is not the case" serves to negate the assertion, "No Democrat is in favor of the Iraq War." In this case, the truth of (10) serves as a warrant for the negation asserted by (11) and so "No Democrat is in favor of the Iraq War" is false. Finally, *if there is no warrant for an assertion or its negation, then it is undetermined in truth.*

Consider

(12) All citizens are taxpayers.

If we find a case in which there is a citizen who is not a taxpayer, for example, someone with no money at all, then the case warrants the negation of (12), that is, "It is not the case that all citizens are taxpayers." In this case, (12) is false and its negation is true.

Consider

(13) Some moon rocks are made of cheese.

In order to warrant the assertion, we would need to find a statement, thing, or event that led to the conclusion that at least one moon rock was made of cheese. In order to find (13) false, we need to locate something to warrant its negation, "It is not that case that some moon rocks are made of cheese" which can be rewritten in standard form, "No moon rocks are made of cheese." Lacking warrant for either conclusion, the statement would be undetermined.

Clearly, from the perspective of inquiry, once assertions have established "truth value" (i.e., they have been warranted true or false), we can also know things about the network of related propositions involving the same terms. The result is a kind of map of connections and divisions among the terms (i.e., classes or collections) and so also among things and events. Suppose that "All citizens are eligible to vote" is warranted and consider the following table,

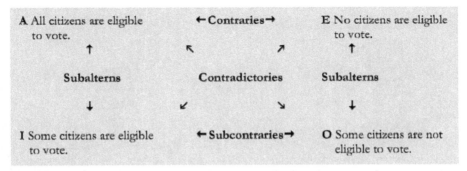

Figure 4.2   Traditional or "Existential" Square of Opposition

usually called "*the square of opposition*" (Fig. 4.2). If the A assertion is true, then its truth warrants the conclusion that the related O assertion is false. That is, if it is warranted that all citizens are eligible to vote, then it must not be true that some citizens are not eligible to vote. The relation between A and O assertions is called *contradictory*. Two assertions are contradictory if one must be true and the other must be false. E and I assertions are also contradictory.

If assertion A is true, the corresponding E proposition is false. But notice that, if A is not true, E is undetermined. This relation, called *contrary*, holds for two statements if both can be false but only one can be true.

If the A assertion is true and *assuming there is something in the class of the subject term*, then it follows that the corresponding I sentence is also true. Notice that if the I assertion is true (so some citizens are eligible to vote), it leaves the A assertion undetermined. This is the *subaltern* relation and it holds when (given the assumption stated), if the universal is true then the particular is true, and when the particular is true, the universal is undetermined.

For Aristotle, syllogisms emerged out of what there is. To make claims about trees or people was to make claims about things in the world. Later in the history of logic, ontology was separated from discussions of logic so that assertions could be treated as strictly abstract. The result changed the meaning of the square of opposition as illustrated here. If an A assertion were about nothing at all, then it could not follow that the corresponding I assertion was necessarily true. The only relation that holds in this so-called "Modern" interpretation of the square of opposition is the relation of contradiction. Even if an A assertion refers to nothing at all, if it is true, then the corresponding O assertion must be false and vice versa (see figure 4.7).

We can also express the standard form assertions graphically. In each case, if an area of a circle is marked out (darkened as in the example below), then there is nothing in it. If it is left open (unmarked), then it indicates that whatever there is in the class or group, it is in the open region. If the area is marked only by an 'x' then the area contains at least one thing. Figures 4.3–4.6 illustrate the graphical representation of each of the four standard

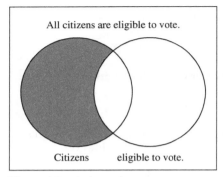

Figure 4.3   A Assertion

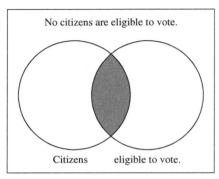

Figure 4.4   E Assertion

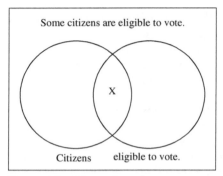

Figure 4.5   I Assertion

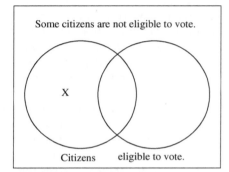

Figure 4.6   O Assertion

form assertions using the subject term of citizens and the predicate term of those eligible to vote.

THE MODERN SQUARE OF OPPOSITION

In the 20th century, under the influence of set theory in which classes could contain no members, relations such as the subaltern relation, where if the A assertion were true then the corresponding I assertion was also true, were rejected. If an A assertion is true, it would not, under this new interpretation, guarantee that the subject of the assertion would have any members at all. For example, it may be true that "All corporations are taxpayers," but this does not guarantee that "Some corporations are taxpayers" is true. Remember that the I assertion requires that there is something in the class of the subject term, in this case, corporations. The A assertion simply claims that if there turn out to be corporations, they will be taxpayers. By itself the claim does not prove that there are such things.

If, for example, one is writing legislation establishing corporations in the first place, there may be a time when, while it is true that corporations will pay taxes, there would be no corporations and so the claim that some corporations are taxpayers would be false. At the same time, if the A assertion is true, the O assertion must be false. If there are corporations (in this new world under legislation), then it cannot be the case that there are some corporations that are not taxpayers. Similarly, if there are some corporations already created that are taxpayers, it cannot be the case that no corporations are taxpayers (the E assertion) and so on. As a result of this change in the underlying metaphysics of the syllogism—that is, the expectation that there can exist classes without members—a new (or revised) square of opposition (Figure 4.7) can be proposed in which the only relation among the terms is that of contradiction.

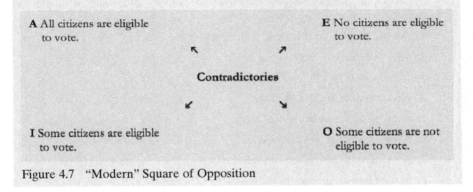

A All citizens are eligible to vote.

E No citizens are eligible to vote.

Contradictories

I Some citizens are eligible to vote.

O Some citizens are not eligible to vote.

Figure 4.7 "Modern" Square of Opposition

4.4 EXERCISES

4.4.1. Rewrite the following statements in standard form:

1. Needed are a couple of good ideas to help get this off the ground.
2. If this group is a cult, then it is not worth joining.
3. Baseball fans cannot watch the World Series unless they have cable.
4. If it's a real coin, then it will not be rejected.
5. Philosophers are the only academics who care about the Good.
6. There is a hair in my soup.
7. Being a fast runner is not a quality shared by all professional athletes.
8. At least this book was worth reading.
9. Not everyone who votes is going to vote for a major party candidate.
10. Nothing is meaningful about vacuous statements.
11. There is no taxation without representation.
12. Not just any random noise can be considered music.

13. Buildings contaminated with asbestos present a serious cancer risk.

14. Snakes can be non-poisonous things.

15. Whoever wrote this song had a big impact on everyone here.

16. Any soldier can find herself or himself in mortal danger.

17. Many patriotic Americans do not vote in presidential elections.

18. Not a single fan showed up for Monday's game.

19. Many flowers bloomed last week.

20. There have been many Republican presidents, G. W. Bush is one of them.

4.4.2. In light of the traditional or existential square of opposition, give the subaltern, contrary, and contradictory of each assertion and its truth value, assuming the original assertion is true.

1. All the players on the basketball team are over six feet tall.

2. All Neanderthals walked upright.

3. No Martians are things that exist.

4. No comedian on SNL is funny.

5. Some proathletes are out of shape.

6. Some politicians are honorable.

7. Some famous artists are not talented.

8. Some teachers are not friendly.

4.5 DIRECT INFERENCE

Standard form assertions give rise to several "direct" forms of inference that, in practice, follow without much reflection (see Fig. 4.14). Consider the claim

(1) No atheists are soldiers.

The relation expressed between the two groups (atheists and soldiers) is such that if the claim is true, then the overlapping category of atheists who are soldiers will have no members at all. In this case, if the first claim is true, then it is also true that

(2) No soldiers are atheists.

This is easily shown using a Venn diagram of the original claim (which shows that the overlapping region between soldiers and atheists has been marked out to indicate that there is nothing in it) (Figs. 4.8 and 4.9).

Since the sequence of the regions in the diagrams is decided by convention and is not necessary in representing the relation between the classes, the same diagram represents both assertions. Given that the diagrams of the two assertions are the same, the two assertions are equivalent. Given the first, one can directly infer the second. Notice also that if the first is true, the second

No atheists are soldiers.

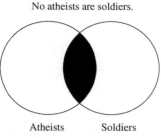

Atheists   Soldiers

Figure 4.8   E Assertion

No soldiers are atheists.

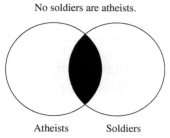

Atheists   Soldiers

Figure 4.9   E Conversion

must be true and so the two assertions validly follow from each other. The relationship between the first E assertion and the second one is called *conversion*. In general, given any E assertion, one can also infer the converted E assertion. If one is true, then other must be as well.

Consider the following A assertions:

(3) All protesters are terrorists.

(4) All terrorists are protesters.

In this case, (4) does not seem related to (3) in the way that (2) was related to (1). Assertion (3) asserts that all protesters are also terrorists but allows that some terrorists might not be protesters. (4) asserts that every terrorist is also a protester but leaves open the possibility that some protesters are not terrorists. Given that two assertions assert different claims (as illustrated by their different Venn diagrams), it is clear that a simple conversion like that used for E assertions will not yield a valid claim. However, if we are given (3), it does follow that

(5) Some terrorists are protesters.

In this case, the resulting I assertion claims that if all protesters are terrorists, AND assuming the existential interpretation of propositions that no class is empty, it would follow directly that at least some terrorists are protesters. This leaves open the possibility that all terrorists are protestors, but does not require it (as does (4)). A assertions can also be converted, that is, the subject and predicate may be switched, but the quantity of the proposition also must be changed from universal to particular. This is called *conversion by limitation*.

It is also possible to convert I assertions. Consider "Some apples are green." In this case, the assertion is that there is something that is a member of both the class of apples and the class of green things. If this is true, then it also follows that "Some green things are apples." However, an O assertion, "Some apples are not green," cannot be converted. Here, the claim is that there is something (an apple) that does not overlap with the other class (of green things). The assertion, if it is true, guarantees that there are some apples that are not green, but it makes no claim about the existence of any green thing at all.

Now consider the following:

(6) All friends are sympathetic.

(7) No friends are non-sympathetic.

The relation of (6) to (7) is the *obverse* relation. Formally, an obverse is formed by changing the quality (not the quantity) of a proposition and then changing the predicate term to its *complement* or opposite. In this case, (6) asserts that all friends will be included in the class of sympathetic people and so its Venn diagram will show that there is nothing in the "friends" region outside the region of sympathetic people. (7) claims that there are no friends that are not included in the class of sympathetic people, that is, though expressed negatively, (7) claims the same state of affairs claimed by (6).

All standard form propositions have equivalent obverses. If "No voters are sixteen-year-olds" then its obverse is "All voters are non-sixteen-year-olds." Notice that the obverse must be true if the original sentence is true and says only that there are no sixteen-year-old voters. The proposition, however, does not rule out there being fifteen-year-old voters or seventeen-year-old voters. Consider the I assertion "Some campaign workers are women" and its obverse: "Some campaign workers are not non-women." In this case, the quality of the original proposition is changed from an affirmative I assertion to a negative O assertion and the predicate term is replaced by its complement: non-women. The resulting phrase, 'not non-women', though cumbersome, means the same thing as the original predicate, 'women'. (This operation will later be called the "double negation" rule that allows one to drop pairs of negations where the first negates the second.) Finally, an O assertion, "Some voters are not racists," has as its obverse "Some voters are non-racists." The predicate phrase of the original is 'not racists' and the predicate phrase of the obverse is 'non-racists' and so means the same thing.

A third form of direct inference is *contraposition*. Consider the following pair of propositions:

(8) All conservatives are principled.

(9) All non-principled people are non-conservatives.

The first proposition holds that all conservatives will be found in the overlapping region so that every conservative is also in the group of principled people. (9) replaces the subject term with the complement of the predicate and replaces the predicate with the complement of the subject. The resulting proposition asserts that anything that is not principled is also not conservative. In a diagram of (8), the only conservatives are within the region of principled people. (9) asserts that if anyone is found to be in the region of the unprincipled, then they are also non-conservative, that is, we will not find any unprincipled conservatives, and so (9) describes the same diagram as the one used to represent (8) (Fig. 4.10).

All conservatives are principled.

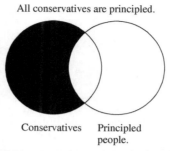

Conservatives   Principled
people.

Figure 4.10   A Assertion and Contrapositive

No politicians are honest.

Politicians   Honest
people.

Figure 4.11   E Assertion

Some dishonest people are not non-politicians.

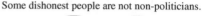
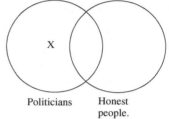

Politicians   Honest
people.

Figure 4.12   E Contrapositive

Consider the E assertion

(10) No politicians are honest.

If we apply contraposition to (10), the resulting assertion "No dishonest people are non-politicians" may not be true. In this case, the second claim is that there are no dishonest people among people who are not politicians. While this may be true (it seems unlikely), it is not guaranteed by the truth of (10). However, if we limit the E assertion to its corresponding O assertion we get

(11) Some dishonest people are not non-politicians.

In this case, assuming that (10) is true and that there are some politicians, then (11) claims that there are some "not-non" politicians in the region outside the region of honest people. The two assertions, (10) and (11), are graphed in Figures 4.11 and 4.12. The "x" in the region marked politicians affirms that there is something in the class. The contrapositive of an E assertion is an O assertion by limitation.

In the case of I assertions, to construct a contrapositive from "Some honest people are politicians" leads to the assertion "Some non-politicians are dishonest." In this case, the first assertion claims that there is something that is both a politician and honest while the second makes the very different claim that there is something that is both a non-politician and dishonest. Regardless of the I assertion selected, no valid contrapositive can be constructed.

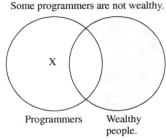

Some programmers are not wealthy.

Programmers   Wealthy people.

Figure 4.13   O Assertion and Contrapositive

However, O assertions do have valid contrapositives. Consider

(12) Some programmers are not wealthy.

(13) Some non-wealthy people are not non-programmers.

(12) asserts that there are people who are both programmers and not wealthy. (13) asserts that there are people outside the region of wealthy people who are "not non-programmers," that is, are programmers (Fig. 4.13).

DIRECT INFERENCE RULES

The following rules assume the existential interpretations of standard form propositions and provide for direct valid inferences. Figure 4.14 summarizes the application of these rules.

Direct inference	A assertions	E assertions	I assertions	O assertions
Original assertion	All citizens are voters.	No citizens are voters.	Some citizens are voters.	Some citizens are not voters.
Converse	(Limitation) Some voters are citizens.	No voters are citizens.	Some voters are citizens.	None
Obverse	No citizens are non-voters.	All citizens are non-voters.	Some citizens are not non-voters.	Some citizens are non-voters.
Contraposition	All non-voters are non-citizens.	(Limitation) Some non-voters are not non-citizens.	None	Some non-voters are not non-citizens.

Figure 4.14   Table of Direct Inferences

Converse: reverse the subject and predicate terms.

Obverse: change the quality (not the quantity) of a proposition and change the predicate term to its *complement* or opposite.

Contraposition: replace the subject term with the complement of the predicate term and the predicate term with the complement of the subject term.

Since direct inferences must be true if the original assertion is true, direct inferences are valid. One can demonstrate that an argument is valid by showing that the conclusion will follow from the original assertion by a series of direct inferences. Consider the following argument that proves, by using the obverse and converse rules, that (4) follows from (1) by a valid argument.

(1) All citizens are voters. (Premise)

(2) No voters are non-citizens. (Obverse of (1))

(3) No non-citizens are voters. (Converse of (2))

(4) All non-citizens are non-voters. (Obverse of (3))

4.6   EXERCISES

4.6.1. Take the standard form propositions from Exercises 4.4.1 and give their direct inferences (converse, obverse, and contrapositive).

4.6.2. From the given premise use direct inferences to arrive at the conclusion.

1. Premise: No smokers are healthy people.
   Conclusion: All non-healthy people are non-non-smokers.

2. Premise: Some trans-fatty foods are not edible.
   Conclusion: Some non-non-trans-fatty foods are non-edible.

3. Premise: All socialists are good comrades.
   Conclusion: No socialists are non-non-non-good comrades.

4. Premise: Some bikers are not drivers.
   Conclusion: Some non-drivers are not non-bikers.

5. Premise: Some tasty drinks are sugary beverages.
   Conclusion: Some non-non-sugary beverages are non-non-tasty drinks.

6. Premise: Some friends in the real world are friends on Facebook.
   Conclusion: Some friends on Facebook are non-non-friends in the real world.

7. Premise: All places where one should shop are small businesses.
   Conclusion: No non-non-places where one should shop are non-small businesses.

8. Premise: No avid TV watchers are avid readers.
   Conclusion: Some non-non-non-avid readers are not non-avid TV watchers.

4.7   THE VALIDITY OF SYLLOGISMS

While propositions in the standard form express a relation between two terms, a syllogism expresses a relation between three terms. Recall Aristotle's definition of a *syllogism* as "discourse in which, certain things being stated, something other than what is stated follows of necessity from their being so" (24b15). The key observation in Aristotle's definition is the idea that certain things follow of necessity from "being so." Put in a more formal way, Aristotle claims that the syllogism provides a general form in which premises and a conclusion are related. Further, in some definable cases, the statements are related in such a way that, regardless of the particular content of the statements, if the premises are true, the conclusion is necessarily true. Syllogistic forms that have this character are called *valid syllogisms*. These forms are especially important because they provide a structured relation that seems to hold in every case and can serve both as a way to understand the relations among things and events as well as between assertions.

Since Aristotle limited his treatment of assertions to those of the standard form, the arguments he considers are composed of assertions that have to do with relations among classes or groups. These assertions are called *categorical assertions* and the resulting arguments are called *categorical syllogisms*.

Consider the following syllogism:

(1) All mammals are animals.

(2) All human beings are mammals.

(3) So, all human beings are animals.

Here, three assertions are presented involving three distinct terms, human beings, animals, and mammals, and each assertion is a universal affirmative. In every argument, the first two propositions are the *premises* and the third is the *conclusion*. Traditionally, the subject of the conclusion is called the *minor term* of the argument, the predicate of the conclusion is called the *major term*, and the term that does not appear in the conclusion is called the *middle term*. The premise that includes the minor term is called the *minor premise* and the premise that includes the major term is the *major premise*. Consider a second example:

(4) All mammals are animals.

(5) No chickens are mammals.

(6) So, all chickens are animals.

In this case, there are again three terms among the three assertions so that the two syllogisms above in fact share a common arrangement of terms. The arrangement of terms for both arguments can be represented by the chart below, where 'S' stands for the minor term, 'P' for the major term, and 'M' for the middle term. In fact, given that every argument includes two premises and a conclusion and exactly three terms, every argument will have one of the

	First figure	Second figure	Third figure	Fourth figure
Major premise	M—P	P—M	M—P	P—M
Minor premise	S—M	S—M	M—S	M—S
Conclusion	S—P	S—P	S—P	S—P

Figure 4.15   Table of Figures of the Syllogism

following forms. These forms (the arrangements of the terms) are called the *figures* of the syllogism (Fig. 4.15). Notice that the pattern of 'M's provides an easy way to recognize each figure.

Consider the argument including assertions (4), (5), and (6) above. While it is in the first figure, the pattern of assertions is different from the first syllogism ((1), (2), and (3) above). In the first syllogism, (1) is A (i.e., a universal affirmative assertion), (2) is A, and (3) is A and so can be represented as having the *mood*, AAA, where *mood* is represented by the sequence of letters naming the form of each assertion of the given argument. The mood of the second syllogism is AEA, where the minor premise is a universal negative.

Once the structure of argument has been formalized as Aristotle proposed, we can begin to consider in detail the nature of formal validity. Since not every mood in every figure gives an argument form that is valid, part of the work of logicians in the history of philosophy has been to identify which of the moods in each of the figures are valid argument forms. This can be done in a variety of ways. We will consider two. The first approach is based on descriptions of the structural differences among the various moods and figures of the syllogism. The second approach is "graphic" and based on the method developed by John Venn in the 1880s. The first approach, which we will call the *axiomatic approach*, considers validity from the perspective of rules that govern how valid syllogisms are organized. One version of this approach is given by Cohen and Nagel. They propose starting with five axioms that govern the quantity and quality of assertions in valid syllogisms.

Axioms of Quantity

   1. The middle term must be distributed at least once.
   2. No term may be distributed in the conclusion which is not distributed in the premises.

Axioms of Quality

   3. If both premises are negative, there is no conclusion.
   4. If one premise is negative, the conclusion must be negative.
   5. If neither premise is negative, the conclusion must be affirmative. (1934, 79)

Using these axioms, we can begin to identify further characteristics of valid syllogisms. Consider, for example, a syllogism of the second figure in the mood

AAA. In this case, the major premise will be of the form "All P are M" and the minor premise will be of the form "All S are M." Notice that the middle term is undistributed in both premises (since A assertions distribute the subject term and not the predicate term) and so the form violates the first axiom. Therefore, the form second figure, mood AAA is invalid. This can be illustrated by an example. Consider:

(7) All Northwesterners are people who live west of the Rockies.

(8) All Oregonians are people who live west of the Rockies.

(9) So, all Oregonians are Northwesterners.

In this case, each assertion is true, but the argument itself is invalid. For if it were valid, it would be necessary that any argument in this form would be such that if the premises are true, the conclusion must also be true. But notice what happens when we modify the terms while keeping the same form:

(10) All Northwesterners are people who pay taxes.

(11) All Illinoisians are people who pay taxes.

(12) So, all Illinoisians are Northwesterners.

In this modified example, the premises are still true, but the conclusion is not, affirming the conclusion based on the axioms that syllogisms in this form are invalid.

The axiomatic approach allows us to produce theorems that follow from the given axioms and provide further rules governing which forms of the syllogism are valid. A *theorem* is an assertion that follows validly from a set of axioms, in this case, the axioms of quality and quantity. So, for example, the following is a theorem:

> The number of distributed terms in the conclusion must be at least one [fewer] than the total number of distributed terms in the premises. (Cohen and Nagel, 1934, 80)

The theorem can be proved by explaining which of the axioms the assertion follows from and how. So

> *Proof:* From Axiom 2, if a term is distributed in the conclusion, it must be distributed in the premises. So, if both terms in the conclusion are distributed, then two terms in the premises must be distributed. But according to Axiom 1, the middle term must also be distributed. So, if both terms in the conclusion are distributed, then three terms in the premises must be. If only one term is distributed in the conclusion, then at least two terms must be distributed in the premises (including the middle term). If no terms are distributed in the conclusion, then at least one must be distributed in the premises (the middle term). Therefore, the number of distributed terms in the conclusion must be at least one fewer than the total number of distributed terms.

Beyond simply confirming individual argument forms as valid or invalid, the axiomatic approach provides for a series of assertions that generalize the notion of validity across all of the possible forms.

In addition to the five axioms of the syllogism given above, Cohen and Nagel also give four general theorems of the syllogism. These are as follows:

i. The number of distributed terms in the conclusion must be at least one less than the total number of distributed terms in the premises.

ii. If both premises are particular, there is no conclusion.

iii. If one premise is particular, the conclusion must be particular.

iv. If the major premise is an affirmative particular and the minor is a negative universal, there is no conclusion. (1934, 80–1)

Using the axioms and these theorems, we can determine which moods will be valid in one or more of the figures. To do this, we can list all of the possible combinations of premises as follows:

AA	EA	IA	OA
AE	EE	IE	OE
AI	EI	IO	OI
AO	EO	II	OO

We can then use the axioms and theorems to eliminate all but eight combinations. For example, using Axiom 3 (If both premises are negative, then there is no conclusion), it follows that we can eliminate the combinations EE, OE, EO, and OO. Using Axiom 1, we can eliminate II. Using Theorem ii, IO and OI can be eliminated and using Theorem iv, IE can be eliminated. The combinations of premises that are valid in at least one figure are AA, EA, IA, OA, AE, AI, EI, and AO.

In addition to the theorems of all figures, there are also special theorems that apply to each figure. They are as follows:

Theorems of the First Figure:

Theorem I: The minor premise must be affirmative.

Theorem II: The major premise must be universal. (Cohen and Nagel, 1934, 84)

Theorems of the Second Figure:

Theorem I: The premises must differ in quality.

Theorem II: The major premise must be universal. (1934, 85)

Theorems of the Third Figure:

Theorem I: The minor premise must be affirmative.

Theorem II: The conclusion must be particular. (1934, 86)

Theorems of the Fourth Figure:

Theorem I: If the major premise is affirmative, the minor premise is universal.

Theorem II: If either premise is negative, the major premise is universal. (1934, 86–7)

PROVING THE THEOREMS

Each of these theorems can be proven using the Axioms and Theorems of all four figures. For example, we can prove Theorem I of the First Figure this way:

Suppose the minor premise is E or O (note that if the minor premise is neither E nor O, then it must be A or I and so affirmative). Axiom 4 says that if a premise is negative, the conclusion must be negative. When this is the case, the predicate of the conclusion must be distributed (the predicates of E and O are distributed). Axiom 2 says that if a term is distributed in the conclusion, it must be distributed in the premises. So P must be distributed in the major premise. Given the combinations of premises that are valid in at least one figure (above) when the minor premise is negative (AE and AO), the major premise must be A and so P is undistributed. Therefore, syllogisms in the first figure where the premises are AE or AO cannot be valid since the predicate of the conclusion cannot be both distributed and undistributed at the same time. All the remaining forms have affirmative minor premises and so the theorem, the minor premises must be affirmative, is proven.

Another approach to determining the validity of forms is to use Venn's method. In this case, validity is illustrated by considering the ontology of the terms asserted in the syllogism. The *graphical approach* allows one to "illustrate" the relations between things or events based on how they are sorted (either on the basis of their properties or on the basis of some other sorting principle). As discussed above, the graphical representation of an assertion, "All citizens are eligible to vote," allows us to imagine a region where we will find all of the citizens and all of those eligible to vote. In this case, the region containing all of the citizens is completely within the region containing those eligible to vote. In order to determine whether or not a syllogistic form is valid, *we simply need to represent the three terms asserted in the premises and see if the resulting picture is correctly described by the conclusion of the syllogism.*

To test a form, begin by stating the form. For example, consider the first figure in the mood AAA. Begin by drawing three interlocking circles and label them S, P, and M (Fig. 4.16). The circles represent the minor (S), major (P), and middle (M) terms. Since each assertion represents the relationship between two terms, each of the assertions of the syllogism can be represented using the models given above of A, E, I, and O assertions. Each assertion can then be represented by shading out some regions (to indicate that there is nothing there), leaving them blank to indicate a distributed term, and placing an X in a region to indicate an undistributed term. Notice that, since the order of graphing the premises does not matter, *either premise can be graphed first unless one is universal and the other is particular. In this case, always graph the universal premise first.* To test the first figure, AAA, graph one of the premises

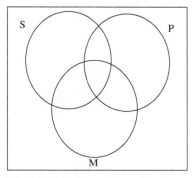

Figure 4.16   Three-Term Venn
Diagram

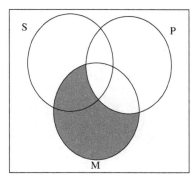

Figure 4.17   "All M are P."

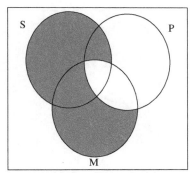

Figure 4.18   "All M are P." and
"All S are M."

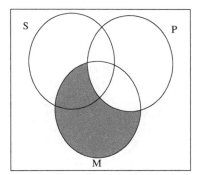

Figure 4.19   "All M are P."

(in this case, the major premise), "All M are P" (Fig. 4.17). The diagram now indicates that all the things that are M are also P, and so the regions of M outside of P are shaded out. Next, graph the other premise, "All S are M" on the *same graph* (Fig. 4.18). Since the minor premise asserts that all things that are S are also M, the graph now shows that all areas of S outside of M are also shaded out. *The syllogism is valid if and only if the graph that remains already exactly represents the conclusion of the syllogism tested.* (There are some exceptions to this requirement that will be discussed below.) In the graph above, the conclusion, "All S are P," is already represented since the only remaining unshaded area of S is also completely within P. Syllogisms in the form of the first figure in the mood AAA are valid.

Next, consider the example given above in propositions (4), (5), and (6) in the first figure and the mood AEA. First, graph the major premise, "All M are P" (Fig. 4.19). Next, graph the minor premise, "No S are M," on the same graph (Fig. 4.20).

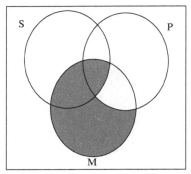

Figure 4.20 "All M are P." and "No S are M."

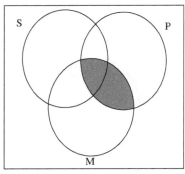

Figure 4.21 "No M are P."

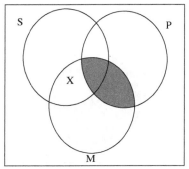

Figure 4.22 "Some S are M."

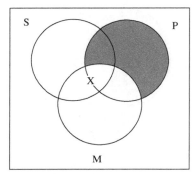

Figure 4.23 "All P are M." and "Some S are M."

The conclusion of the argument is "All S are P." Check the graph. The argument is valid if and only if the graph already exactly represents the situation where all things S are also P. The graph *does not* represent this situation because there is a region of S that is outside of P. Therefore, syllogisms of the first figure and the mood AEA are invalid.

Now, consider syllogisms of the fourth figure in the mood EIO. In this case, since one premise is particular, graph the universal assertion, "No M are P," first (Fig. 4.21). Next, graph the minor premise, "Some S are M," on the same graph (Fig. 4.22).

The conclusion, "Some S are not P," is shown by having some S in a region outside the region of P and the graph above shows this to be the case. An X indicating that there is at least something in the region where S and M overlap means that there is in fact some S that is not P.

Next, consider syllogisms of the second figure in the mood AIO. The graph of the minor and major premises is shown in Figure 4.23.

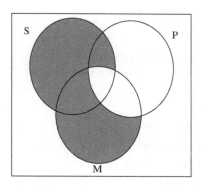

Figure 4.24 "All M are P." and "All S are M."

In this case, placing the X on the line between S and P indicates that the minor premise is ambiguous. The premise reads "Some S are M" and so we should put an X in a region where S and M overlap. But two such regions are available. Which do we choose? If we choose the region outside P, it would appear that the conclusion, "Some S are not P," is represented. But since this doesn't follow necessarily—that is, because the premises are not specific enough—syllogisms in the second figure, mood AIO are invalid.

Finally, consider an argument in the first figure, mood AAI. In this case, when the two premises are graphed, we get the following results shown in Figure 4.24.

The conclusion is an I assertion, "Some S are P." The graph, does not, however, include an X in the region of the graph that includes both S and P. It would appear that the form is not valid. But recall that, for Aristotle, assertions are not abstract and formal, but tied to the real things. When we claim, for example, "All tigers are cats," we assume the existence of at least some tigers and cats. If this is so, then, in the above graph, Aristotle will suppose that there are at least some things that are S and some things that are P. Since there is only one region of S and it is completely enclosed in P, it would follow on this interpretation that there is something that is S and it must also be P. In this case, the argument form is valid. This particular kind of valid form is called "weakly" valid since the conclusion asserts that some S are P when, on the existential interpretation, the stronger claim, that all S are P, is also true.

Practice checking the validity of various forms on your own by graphing the following and determining whether or not they are valid forms. In some cases, the forms will be valid but only under the existential interpretation (and so will be weak forms).

In the first figure, moods AAI, AEO, EAO

In the second figure, moods EAI, AEE, EAE

In the third figure, moods IAI, EAI

In the fourth figure, moods IAI, EAA, AEE, AEO

TRADITIONAL NAMES AND VALIDITY

Traditionally, each valid argument form (figure and mood) had a name that helped logicians remember the various forms. In each case, the vowels in the name give the mood.

First Figure: Barbara, Celarent, Darii, Ferioque

Second Figure: Cesare, Camestres, Festino, Baroco

Third Figure: Darapti, Disamis, Datisi, Felapton, Bocardo, Ferison

Fourth Figure: Bramantip, Camenes, Dimaris, Fesapo, Fresison

Some forms are only valid given the existential interpretation of the square of opposition. These *existentially valid (or weak) forms* are

Bramantip (Fourth Figure)

Fesapo (Fourth Figure)

Darapti (Third Figure)

Felapton (Third Figure)

Finally, three other argument forms are also "weakly" valid (but have no names). In each case, the premises justify the stronger conclusion, that no S are P. Notice that EAO in the first figure is weakly valid and the stronger form, EAE (Celarent) is also valid.

First Figure EAO

Second Figure EAO

Fourth Figure AEO

4.8 EXERCISES

4.8.1. For each syllogism below, rewrite its statements in standard form and then give its figure and mood (e.g., "first figure, mood AAA" or its name, e.g., Barbara). Next, using the Venn diagram method, prove that the syllogism's form is valid or invalid. Show the resulting graph and if the form is not valid, explain why in reference to the Venn diagram you have constructed.

1. Not a single Republican is an advocate of socialized medicine. And yet, everyone in favor of smaller government is also a Republican. I think that it follows, and you will agree, that no one in favor of small government is a supporter of socialized medicine.

2. Everybody who knows their subject well turns out to be a good teacher. A few of those who know their subject well are scientists. So it follows that only some scientists are good teachers.

3. Some of the politicians I have encountered are not excellent musicians. At the same time, every barber I have met has been an excellent musician. So,

it follows that some of the politicians I have met over the years are clearly not barbers.

4.8.2. Give the figure and mood of each of the following and show whether it is valid or invalid.

1. **Premise 1:** All ninjas are good fighters.
   Premise 2: All black-hooded strangers are ninjas.
   Conclusion: All black-hooded strangers are good fighters.

2. **Premise 1:** No human-made thing is everlasting.
   Premise 2: Some of the most beautiful things are human-made things.
   Conclusion: Some of the most beautiful things are not everlasting.

3. **Premise 1:** No coffee shops are places that are quiet.
   Premise 2: All places that are quiet are places that inspire the imagination.
   Conclusion: Some coffee shops are places that inspire the imagination.

4. **Premise 1:** All accountants are better looking than movie stars.
   Premise 2: No lawyers are better looking than movie stars.
   Conclusion: No lawyers are accountants.

5. **Premise 1:** No great thinkers are those who avoid philosophy.
   Premise 2: Some college professors are those who avoid philosophy.
   Conclusion: Some college professors are not great thinkers.

6. **Premise 1:** Some things worth doing are things that can get out of hand.
   Premise 2: Some logic problems are things that can get out of hand.
   Conclusion: Some logic problems are things worth doing.

7. **Premise 1:** Some exercises in this book are easy.
   Premise 2: All exercises in this book are important things to do.
   Conclusion: Some important things to do are easy.

8. **Premise 1:** Some movies are not worth watching.
   Premise 2: All movies are extravagantly made things.
   Conclusion: Some extravagantly made things are not worth watching.

9. **Premise 1:** No well-planned actions are revolutions.
   Premise 2: All well-planned actions are successful enterprises.
   Conclusion: No revolutions are successful enterprises.

10. **Premise 1:** No things worth doing are criminal offenses.
    Premise 2: Some criminal offenses are dangerous actions.
    Conclusion: Some dangerous actions are not things worth doing.

11. **Premise 1:** All church-going people are trustworthy people.
    Premise 2: No trustworthy people are car salesmen.
    Conclusion: No car salesmen are church-going people.

12. **Premise 1:** All those who act only for the money are those who are famous.
    Premise 2: No good movie actors are those who act only for money.
    Conclusion: Some of those who are famous are good movie actors.

CHAPTER 5

INDUCTION AND THE LIMITS OF REASON

5.1   LIMITS OF THE SYLLOGISM

The theory of the syllogism aimed to account for both the laws of thought according to which demonstration is possible and the laws of "being itself," which guaranteed the truth of demonstrations. As logic was reformed in light of work in mathematics in the late 19th century, logicians typically focused on the formal relations expressed in demonstrations and in mathematics in order to devise simpler sets of laws and symbols to capture validity. One philosopher who resisted the trend toward a purely formal understanding of logic was Josiah Royce. Born in Grass Valley, California, in 1855, Royce became a faculty member at Harvard in the 1880s and in the early 1890s became renowned as the leading American advocate of idealism.

Royce's idealism, in contrast to the realism of the day, argued that things and events can only ever be understood as moments of consciousness, that is, as ideas. One might believe, as a realist, that there are real things beyond ideas that have just the characteristics we know them to have. Things are what they are independently of their being known and it is these things alone that are real (and ideas are at best appearance). Idealists responded that even these supposedly independent real things could only have the character that they have in the context of some experience. So if these real things are anything at all, what they are will still turn out to be a matter of the ideas about them (how they are known in experience). Realists, in short, could only posit reality as an *idea* and so idealism must be the better perspective for understanding the world.

Further, since thought, for the idealists, was a process of judgment (i.e., of judging things to be chairs or tables or people), idealist logic was a theory of judgment in which validity was understood as the principle that held systems of judgments together. The idealist logician Bernard Bosanquet explained, "The forms of thought have the relation which is their truth in their power to constitute a totality; which power, as referred to the individual mind, is its power to understand a totality. The work of intellectually constituting that totality which we call the real world is the work of knowledge" (Bosanquet, 1911, Vol. I, 2–3). The work of logic is "analyzing the process of this constitution or determination." Logic, he continued, "has no criterion of truth nor test of reasoning. Truth is individual, and no general principle, no abstract reflection, can be adequate to the content of what is individual." The consequence of this view is that "The real world for every individual is thus emphatically his world; an extension and determination of his present perception" (1911, Vol. I, 3).

The idealist view of logic raised strong objections at the end of the 19th century from the perspective of science and mathematics, fields committed to the idea that the real world is not the world of an individual, but rather something independent of individuals that can be discovered and known and where logic is about the forms of reasoning that preserve truth, again, independent of individuals, not merely the rules by which a "totality" is understood.[1] By the early 20th century, realism came to dominate the field of logic as the view that could make sense of science and mathematics and, by World War I, idealist logic was set aside.

Royce began to see problems with idealism in the late 1890s and, in 1898 after attending a series of lectures by Peirce, began to rethink the idealist conception of logic. He realized, with Peirce, that whether one were an idealist or a realist, whatever particular things or events came into question—whether they were ideas or "external" objects—particulars can only emerge in experience or in the world as a result of a series of ordered relations. When you set out to go to class, or to write a paper, or to call a friend, the events unfold in a sequence without which the particular activities would make little sense. The series of interactions that lead to a phone conversation with a friend gain their particular character in the context of a wider purpose, perhaps to find out the

---

[1] Ultimately, the idea that a totality could be understood was rejected by Gottlob Frege and Bertrand Russell in light of the so-called "Russell paradox." In the face of the paradox, Frege gave up logic while Russell went on to develop the "theory of types" which, he argued, escaped the paradox by proposing that for every logical language, there was a wider logical language—a meta-language—which encompassed it. The theory of types ran into problems as well.

assignment given in class or to plan to go to a game on campus. This same conversation can give rise to new purposes that order still more activities that might lead to a study session or a meeting for coffee.

For Royce, the study of order could make sense of realism and idealism and in a way that could stand as a more general theory accounting for the possibility of particular events and particular inferences in both philosophical perspectives. "Order," Royce says, "is known to us through inference; i.e. the orderly is that which corresponds, in the real or the ideal world, to what we infer when we systematically draw conclusion from premises. There the understanding of the inmost nature of order logically depends upon understanding the relations on which our power to infer rests" (Royce, 1913, 230).

Consistent with the idealist project of logic as the study of judgments—of the will and its activities—Royce sees logic as an effort to understand types of order that are produced by judgments. At the same time, consistent with the realist project, Royce took logic to be a study of the real order systems that constrain judgment and frame truth. Logic, then, can be understood as a "subordinate part" of "the science of order." The science of order, he claims, "is indeed incidentally concerned with the norms of the thinking process," that is, given its interest in the principles of order, it is also a resource for learning to think well and to criticize poor thinking.

In the end, however, the science of order aims to provide insight into the structure of interactions that include thinking but also include the general conditions of what Royce calls "rational action" but is probably better called "purposive action."[2] In this sense, the science of order is about the structure in which action can take place, that is, the structure of agency. Agency is usually taken to mean the power to act. The science of order is the study of the most general principles affecting the possibility of taking action and so provides the context for human interaction—with each other, with non-human others, technologies and so on. For Royce, the science of order becomes the framework for logic as well as metaphysics, aesthetics, ethics, and social philosophy.

Royce begins his account of the principles of logic with a discussion of "methodology," that is, the practices involved in the process of thinking. For Royce, thinking has the unique property of being a practice that creates and guides, but also one that has the potential to reflect upon itself in order to modify and improve its own practice. "Everywhere," he says, "the consciousness of method grows in proportion as thought comes to play a successful part in the organization of human life" (1914, 311). As a result, Royce continues, methodology is "the mother of Logic" in that reflection on its own practices leads beyond a systematic and critical examination of the practices of thinking

---

[2] While Royce uses "rational" in his discussion of modes of action, his work in *The World and the Individual* suggests that "rational" could also be taken as equivalent to "purposive." "At any moment your ideas, in so far as they are rational, embody a purpose" (1899, 441). Also see *Lectures on Modern Idealism*: "The world of our experience becomes a rational realm to us in so far as we can interpret it in terms of ideas that adequately express our own conscious purpose" (1919, 245).

to "the *Forms*, the *Categories*, the *Types of Order*, which characterize any realm of objects" (Royce, 1914, 312).

Aristotle's logic, on this account, can be seen as one of the first attempts in what we now call the Western philosophical tradition to understand types of order. As we have already observed, Aristotle saw the theory of the syllogism as grounded in an ontological perspective that takes the categories of thought as extensions of the categories that structure the world. While Aristotle could recognize that our thinking can be wrong despite its connections with the world, he also held that we can be right in our understanding because what we know (when our knowing is done carefully and correctly) emerges from the things and events that we know. Royce summarizes: "orderliness and system are much the same in their most general characters, whether they appear in a Platonic dialogue, or in a modern text-book of botany, or in the commercial conduct of a business firm, or in the arrangement and discipline of an army, or in a legal code, or in a work of art, or even in a dance or in the planning of a dinner. Order is order. System is system. . . . Whatever one's metaphysic may be, one must therefore recognize that there is something objective about the Order both of our thought, and of the things concerning which we think" (Royce, 1914, 316).

There are two common approaches to defining the categories that serve to order syllogistic reasoning: definition and enumeration. The first method, *definition*, involves the process of identifying characteristics of the things and events considered in order to generate descriptions of the features common to all members of the category. If we are interested in studying trees of the Northwest, we begin by collecting samples and then organizing them in light of the features they display. Eventually we can generate definitions of various kinds of trees. A Western red cedar (*Thuja plicata*), for example, can be defined as a "Large to very large tree with tapering trunk, buttressed at base, and a narrow, conical crown of short, spreading branches drooping at the ends: foliage is resinous and aromatic" (Whitney, 1985, 368). We can go on to describe the leaves, bark, twigs and cones of the tree. The method is comparative in practice since we can determine different groups by directly comparing features of the samples and we can identify new instances of defined categories by comparing new samples with the existing definitions.

Aristotle's conception of how we determine kinds and classes is an elaborate process of sorting features (and so kinds) hierarchically so that all things are related through larger and larger categories, but each kind (such as Western red cedars) is also distinct in the larger system of order. In general, for Aristotle, classification could always be understood in terms of a genus/species relationship in which the species was contained within the genus (essentially the same relation we find in standard form assertions). When classifying species, Aristotle's definitions were given in terms of genus and "differentia." For example, the genus might be "bird" and the differentia might be that some birds feed in the water while others eat berries on land. At the same time, the classification of birds would indicate a differentia among the larger genus of animals. While Aristotle's approach to classification dominated until the 18th century, Carolus Linneaus eventually generated a more precise taxonomic system in which all organisms can be classified in terms of their kingdom, phylum, class, order, family, genus, and species (with kingdom as the widest class and species the narrowest classification). The wolf, for example, will be classified within the kingdom *Animalia*, the phylum *Chordata*, the class *mammalian*, the order *Carnivora*, the family *Canidea*, the genus *Canis*, and the species *lupus*.

The second method, *enumeration*, involves listing members of the class or kind in order to see if a particular individual is part of the collection. In order to find out if Chester Arthur is among the presidents of the United States, we can enumerate (list one after another) all of the presidents. Should we discover Chester Arthur on the list, we can conclude that he is (or was) a U.S. president. If he is not on the list, we can conclude that he is not. In this case, we do not try to offer a precise definition, but simply to get all of the instances of the class to present themselves.

As Royce observes, the two methods of definition and enumeration are "in fact, by no means always very sharply to be distinguished." The use of definitions seems to depend upon our ability to list at least some of the members of a given class in order to discover the features that they have in common. Given an established definition, the enumeration we might give for a particular class may begin with a definition that will guide us in selecting the individuals to be enumerated.

Aristotle's theory of the syllogism is framed by definition and enumeration, that is, it begins with the commitment that things in the world are what they are in relation to other things so that we can discover their identity through comparison and enumeration. For Aristotle (and for much of the Western philosophic tradition), logic starts by affirming the certainty of classification "in reality" and secures the possibility of knowledge by definition and enumeration.

There are, however, at least four clear limitations to Aristotle's view. Consider the syllogism:

> All human beings are mortal.
>
> Barack Obama is a human being.
>
> So, Barack Obama is mortal.

The assumptions that we need to make in order for the syllogism to hold include the fact that *there is a class of human beings* and that *they are all, without exception, mortal.* The first difficulty here is that it is hard to say exactly where the class of human beings begins and ends. Can we tell exactly which human being was the first one? Was there ever such a being? Can we tell which human being will be the last? If the major premise is to be true, there must be a complete class of human beings all of whom have the character (even while alive) of dying. We are, of course, willing to hold this premise under most circumstances and willing to say that among this class of human beings is Barack Obama who is likewise mortal. But is this any more than a hypothetical claim that awaits verification?

The second limitation, based on the realization that there are classes that we cannot fully know, leads to the conclusion that the appropriate way to understand a syllogism is as strictly hypothetical. We do not know that any given living human being will die, only that, given our past experience with human beings, most do. As a result, the syllogism in the first figure and the mood AAA seems to be at best a hypothesis. This leads to the so-called "hypothetical" or "Modern" interpretation of the square of opposition discussed in Chapter 3. Since we cannot be certain that there are members of any particular class we choose, or that all the members of the class share the property in question, we can only hypothesize. So instead of "All human beings are mortal," we say "If a thing is a human being, then it is mortal." This assertion does not require that there be a complete class of human beings at all, only a definition awaiting the discovery of something that fits it. The hypothesis is partially confirmed each time one finds a human being who does.

The third limitation for the Aristotelian model of the syllogism is that it fails to take into account changes in the nature of any given class or kind. When evolution was accepted as the primary way to understand the development of organic species, philosophers argued that the first thing to be changed was the long-standing Aristotelian claim that species are neither created nor destroyed. Aristotle, as we mentioned above, had to hold that the classes referred to by the terms of a syllogism (e.g., human beings) must be complete classes. In this case, there is necessarily a determined class, called 'human beings', that always exists and which always serves as the referent for our claims about the world. Evolution theory undermined this ontology by arguing that species change and concluded that there are no fully determinate classes. As a result, again, logicians sought new ways to understand Aristotle's work and to expand the field of logic into a theory not dependent on an ontologically static universe.

Even if we bracket the idea that Darwin undermined the ontology required for Aristotelian logic, the pragmatists Peirce, William James, Dewey, and Royce also argued for reform in ontology in light of a fourth problem: the old Aristotelian model seemed to make classes strictly a matter of discovery.[3] Human beings, or any other class, are already "in" their classes before any inquirer seeks them out. On this view, things in the world are to be found and the role of the inquirer is essentially passive. In the 18th and 19th centuries these commitments led to theories of knowledge that Dewey called "spectator theories" in which the inquirer is an observer who can find things as they are but in "finding" them does not affect them. In lieu of these spectator theories

---

[3] Dewey discusses the limitations of Aristotle's logic in "The Needed Reform of Logic" (Dewey, 1938, Chapter 5). He argues in part that Aristotle makes a "logic of discovery and invention" impossible. Only one form of "discovery" is possible under the Aristotelian model, which is limited by "the apprehension of the species present in objects of perception on one hand and rational grasp of some essence defining a complete species on the other hand." "Learning," under these conditions, "merely brings these two antecedently given forms of knowledge into connection with each other. ... [Invention] of the new has no place. ['Discovery'] had only its literal etymological meaning of *coming upon* something already there" (Dewey, 1938, 92–3).

(whether idealist or realist), the pragmatists adopted an alternative theory of logic and ontology that returned in important ways to Aristotle but reconceived Aristotle's syllogisms as *limit cases of reasoning*. From this perspective, logic still tells us about the relations among things and events, but does not require that any of the kinds and classes be complete. In effect, logic (including the theory of the syllogism) can teach us about systems of order while allowing that our inquiries can proceed without the certainty of a static universe.

5.2   THE PRINCIPLES OF INDUCTION

The method that initiates inquiries in response to disruption and novelty is what Dewey called "suggestion" and what Peirce called "abduction." This form of reasoning is guesswork and the least certain mode of thinking. At the other extreme, deduction is the process identified by Aristotle in the operations of valid syllogisms in which the premises necessitate the conclusion and so is the most certain mode of thinking. Between these two modes with respect to certainty is the mode of induction, which Peirce called *ampliative reasoning*. Induction is the process of following the suggestions made through abduction by trying proposed causes as a means of finding what will bring about the result sought or to be explained. Since this process of trying (or testing) is always confined to a finite number of attempts, its potential for certainty is limited. At the same time, its practice can serve to transform guesses into useful knowledge that can foster understanding and further action.

Induction, then, can be characterized in two ways. First, the assertions (and the things or events denoted) are connected in ways that are less than certain.

> Sarah is usually late for our meetings.
>
> So Sarah will probably be late for our meeting today.

We do not mean that Sarah necessarily will be late, but rather that late arrival is likely. The conclusion is connected to the premises, but the connection is less than certain and the things and events in question are related in a way that may not lead to Sarah's being late this time. From this perspective, even though an inductive argument is an argument (a process of thought that leads to belief proceeding from premises), *inductive argument forms are not deductively valid* (i.e., they are not of a form where if the premises are true than the conclusion is necessarily true).

Inductive arguments can be characterized in a second way as well. As illustrated in the case of Sarah's arrival at a meeting, *inductive arguments extrapolate or extend patterns to new cases*. In cases of deduction, all cases are presumed to be present somehow in the classes that are denoted by the terms of the syllogism. In induction, some range of cases is available but these cases nevertheless leave the classes incomplete. In the case above, the premise states a pattern of behavior covering many cases in which Sarah has been late for meetings. The conclusion extends the cases of Sarah's meeting attendance to one more and applies the established result: she will be late.

Consider a syllogism in the first figure, mood AAA.

> All cats are mammals.
> All tigers are cats.
> So, all tigers are mammals.

The connections among the terms are necessary (both in inference and in reality) if and only if the terms are complete. That is, in order that the conclusion is necessarily true (according to the definition of validity), the classes denoted by the terms must be determined by a complete definition (it defines all and only the features of things that are in the classes). The classes denoted by the terms must also be complete in the sense that all of the individuals within the class must "exist" in some way that blocks the possibility of some future individuals fitting the definition but not being an instance of the class.

The lack of sharp boundaries for classes as a result of evolution (changing classes) and interaction (the role of the inquirer in "making" classes) means that the form of the syllogism is not valid in the strict sense in any particular instance but rather is in the form of *an ideal limit that will be approached through each particular case but which may never actually be achieved.* In effect, the syllogism presents a form that eliminates ambiguity both in the world and in our thinking but does so in a way that anticipates a point at which the subject and predicate terms are actually complete. So long as the class of cats is changing, for example, then there are possible cases in which the premises may be true and the conclusion false. The result is not a mere hypothesis about cats and tigers, but rather a representation of what is really the case now. There are cats and tigers and they are related in a structured way. As more cats and tigers live and die, the more closely the classes approach completion and the connection among the terms of the syllogism approaches a necessary connection.

Nevertheless, so long as classes develop and change and inquirers expand their explanations and adjust their definitions, the point of completion will always be a bit further down the road and the point of completion will remain just out of reach. Notice, however, that even as the point of completion of the classes remains out of reach, the possibility of its completion stands as an *organizing principle* in terms of which the classes remain connected even as they evolve. The relations are, in these cases, not arbitrary just as they are not fully determined. Despite its incompleteness, the syllogism still points the way to what we call deduction and its characteristic of certainty.

But what does it take for an argument regarding incomplete or changing categories to work at all? Suppose, for example, that after careful data collection, a researcher concludes that all youth will vote Democratic in an upcoming election. There are two obvious problems with the assertion, even as a rough guide to understanding voting patterns and planning campaigns.

First, the claim that "all youth" will vote in a certain way clearly cannot mean "all" in the Aristotelian sense. Even assuming that the researcher was able to survey every eligible young voter at the time of her study, thanks to the dynamic character of human life, new voters come of age every day and other voters become ineligible. Exactly how many voters must claim to vote Democratic in order to conclude all (or even most) of them will vote Democratic?

Second, assuming that there is an agreed-upon number of voters that will verify the claim that all (or most) young voters will vote Democratic, the fact that they have voted one way or have said that they will vote one way does not guarantee that they will. Nothing in principle stops every person surveyed from voting a different way in a future election.

The first problem has to do with how the subject category is quantified. What portion of the subject category does it take to allow the assertion that the predicate will apply to all or most of the category? Or, if the claim is that only a portion of the subject category is represented, how is the connection between the subject and predicate expressed?

The second problem is related to the one made famous (philosophically) by David Hume. While it may be possible to state clearly what one has learned directly from observation, Hume argued, events predicted cannot be observed. Instead, we routinely observe events that occur together or successively and then expect that they will occur in the same way in the future. Such reasoning could be certain if these "constantly conjoined" things or events were necessarily connected. Hume concludes, however, that the relation of necessary connection itself can also never be observed so, at best, we can assume the connection despite a lack of observed evidence. In our voter research, the fact that most or all of those surveyed said they would vote one way or another does not guarantee that they will because there is no verifiable necessary connection between what they say and what they do.

The general pattern of reasoning which cannot count on deductive certainty and instead moves from partial knowledge to a wider claim is "ampliative" reasoning, since the conclusion amplifies the evidence beyond what is immediately warranted, through a process of induction. Hume's rejection of the idea of necessary connection undermines induction by making the connection between the facts we have observed and the predictions we make a matter of custom (or habit) at best. Induction can bring no guarantees and so, the argument goes, it appears that induction is little better than the guesswork it is meant to help.

THE PROBLEMS OF INDUCTION

The central problem of induction is that, without necessary connection, any inductive conclusion is subject to failure and so, even with true premises, may lead to a false conclusion. Attempts to address the problem by devising special rules of induction in terms of which inductive arguments can be valid have been frustrated as well.

One approach to securing induction is to establish a special set of rules in terms of which inductive arguments are valid. The goal, as Nelson Goodman proposed it, "is to define the relation that obtains between any statement S_1 and another S_2 if and only if S_1 may properly be said to confirm S_2 in any degree" (1983, 67). So, given some observations about voting, a researcher may then make a more general claim, for example, most young voters will vote Democratic. Now consider the assertion that golden retrievers do not vote. The observation asserts that we have found another group whose members do not vote Democratic and whose members are also not young voters. Since this observation can count as inductive evidence in support of some more general claim, it serves to help confirm (among other things) the claim that those who do not vote Democratic are not young voters. The claim that golden retrievers do not vote must then also amount to evidence that young voters are Democratic voters (by contraposition). While the researcher may appreciate the support, the fact that obviously irrelevant observations can count as legitimate evidence for an inductive conclusion suggests that the process of induction is seriously flawed.

The problem is that the evidence is taken in a context that is not restrictive enough and allows observations about golden retrievers (among many other things) to count as evidence in a discussion focused on voters in the next election. Goodman argues that this can be corrected by requiring that the evidence is restricted to a relevant domain, in this case, eligible (human) voters in appropriate categories (Democratic, Republican, and other affiliations as well as categories of age).

Goodman, however, argues that even this sensible restriction cannot provide the guarantees necessary for validity. To demonstrate this, Goodman proposes what he calls "the new riddle of induction." He proposes the riddle this way: "Suppose that all emeralds examined before a certain time t are green. At time t, then our observations support the hypothesis that all emeralds are green" (1983, 73–4). After viewing a few emeralds, it appears at time t that the hypothesis is correct. He continues: "Now let me introduce another predicate less familiar than 'green.' It is the predicate 'grue' and it applies to all things examined before time t just in case they are green but to other things just in case they are blue" (1983, 74). Since the observations I made before time t are of green emeralds, the observations also support the assertion that all emeralds are grue. Since the predictions implied by the two predicates are different (and contradictory), it appears that my observations will produce a contradiction. The problem is not, Goodman argues, a problem with observation in relation to a hypothesis, but rather a problem with the hypothesis. What is needed is "a way of distinguishing lawlike hypotheses, to which our definition of confirmation applies, from accidental hypotheses, to which it does not" (1983, 82–3). He concludes "the new riddle of induction, which is more broadly the problem of distinguishing between projectible and non-projectible hypotheses, is as important as it is exasperating" (1983, 83).

One of the ways of responding to the problem of induction proposed by Hume is to make two assumptions that govern one's ability to extrapolate from particulars to general conclusions. The first assumption is *"that nature is uniform"* and the second is *"'that every event'* (or as one sometimes asserts, *'every individual fact') 'has its sufficient reason'"* (Royce, 1914, 322).

The first principle is easy to see. Suppose we want to determine something about the trees on a particular acre of land. We begin counting the trees according to their various categories, and conclude that there are 100 trees, exactly 50 are Western red cedar and the other 50 are Douglas firs. Even in a case where we can check each and every one of the members of the class of trees on the given acre, we nevertheless assume that nature is stable and that the red cedars will remain red cedars while the Douglas firs remain Douglas firs. In a case where we explain to someone about to visit our acre of forest that there are 100 trees divided equally between red cedars and Douglas firs, we expect that the conditions that held during our count of the trees continue to hold.

The second assumption requires that things and events that we encounter are what they are for some sufficient reason, that is, "something that, from its nature, is general and capable of being formulated as a law of nature" (Royce, 1914, 323). In this case, new observations can be expected to confirm a given principle *"unless there is some sufficient reason why they should not"* (Royce, 1914, 323). This reason must then be another law of nature. Taken together, these assumptions appear to lay a foundation for determining the validity of inductive arguments by assuming the stability of what is studied and that whatever changes occur will accord with a law.

In his criticism of the received view, Royce argues "Familiar as such modes of stating the warrant for generalizations and extrapolations are, it requires but little reflection to see that the formulations just stated *leave untouched the most important features of the very problem that they propose to solve*" (1914, 323). The principle of the uniformity of nature is not useful since, even though there are many observed uniformities in nature, the principle cannot help an inquirer identify *which* uniformities are important to the investigation at hand. In the case of the forest, while it is certainly clear that the uniformity across time of the trees counted is important, this particular uniformity should be set aside if the investigation regards forest succession. Here, while the boundaries of the forest might be arbitrarily determined and uniform through time, the types and distribution of trees must be viewed as changing through time.

Royce argues that the principle of sufficient reason is "[e]qually useless in aiding us with reference to any one decision regarding our right to generalization and extrapolation" (1914, 324). The commitment to the idea that novel observations must conform to some sufficient reason provides no particular help in determining of what order the observation is a part. "This question can admittedly be answered with certainty only when the now unobserved facts have come to be observed" (1914, 325). Whatever law-like changes may emerge, they do not emerge in advance of observation but are constructed through observation. Even if the principle of sufficient reason applied to the universe, it would provide little help to the limited perspective provided by induction. Neither principle, Royce concludes can provide any guidance for determining the probability that underlies the claim that all or most young voters will vote Democratic or the likelihood with which a particular young voter will vote Democratic. In order to determine these, we may turn to Peirce and his analysis of induction.

In a series of papers on induction written in the late 19th and early 20th centuries, Peirce considered diverse forms of induction in search of a set of principles in terms of which inductive reasoning could be valid. The result, summarized by Royce in his *Principles of Logic*, included three principles that frame inductive arguments of all sorts. The first holds that the situation to which an argument is applied is *finite*, "a finite set of facts" (Royce, 1914, 326). Second, given a finite situation (or set of facts), the facts must have "*some definite constitution*," that is, "there are possible assertions to be made about these facts which are either true or false of each individual fact in the set in question" (1914, 326). And third, that the argument is based on a *random (or fair) sample* of the facts or terms within the bounded situation.

The third principle distinguishes cases of induction from deduction in the Aristotelian sense. If the entire class of facts or terms were available to the argument, then a sample would not be required. Since the entire class is not available, a sample chosen at random will provide something like a limited model of the class as a whole. The second principle observes that the process of selection requires that the situation include terms that are discrete and so may be selected to represent the whole. "Thus," Royce says, "if our realm of 'objects of possible experience' [the situation at hand] is a realm wherein men may be conceived to be present, and if the term *man* has a precise meaning, then the assertion, made of any object A in that realm, 'A is a man,' is either true or not true of A" (1914, 326).

Finally, if the situation is not finitely bounded, a random selection would be impossible. If, for example, one tries to select randomly from a bottomless bag of marbles, whatever sample is selected, no matter how large, will always represent an infinitely small sample of the whole. While one may make assertions about the whole on the basis of these samples, the resulting claims will be guesses alone and will not admit of any degree of probability. As Royce concludes on Peirce's behalf, "given a finite collection of facts which has any determinate constitution whatever—be this constitution more or less 'uniform,' be the 'sufficient reason' for this constitution some one law, or any possible aggregate of heterogeneous 'reasons' whatever—it remains true that we can, with probability, although, of course, only with probability, judge the constitution of the whole collection by the constitution of parts which are 'fair samples' of that whole, even when the collection is very large and the samples comparatively small" (1914, 327).

Inductive arguments come in a variety of forms. Arguments about the probability of something occurring given something else have a deductive character based on a determination of the relation of the things or events in question. These include generalizations, analogical arguments, arguments using John Stuart Mill's methods of similarity, difference, residue, and concomitant variation, and probability logic.

In general, inductive arguments present a limited range of information that, if properly collected and used, can support conclusions (beyond guesswork) in the practice either of investigation or making an argument. The three conditions just discussed provide a general framework for proper collection and use of information (or other resources of inquiry) and form the basis for inductive validity. The first requires that the situation in which the inquiry or argument occurs have a boundary or horizon (in Habermas's terms), whether this is provided explicitly in the process of investigation or argument or is left as part of the background. The second requires that "facts" of an investigation or argument are "determinate" (or "predesignated," as Peirce explains it).[4] Together these two conditions stabilize the dynamic character of experience and set the stage for testing hypotheses that will lead to amplified conclusions.

The third condition stipulates a general standard for how one may try a hypothesis. The idea of a "fair sample" is applied to a certain range of investigations and arguments, but it also marks a disposition or approach taken in the process of fair sampling that also underlies others forms of inductive argument including those using analogy or causation as a means of testing and arguing for amplified conclusions. This condition of validity is, in its broadest form, comparable to what Dewey calls the "scientific attitude":

> On its negative side, it is freedom from control by routine, prejudice, dogma, unexamined tradition, sheer self-interest. Positively, it is the will to inquire, to examine, to draw conclusions only on the basis of evidence after taking pains to gather all available evidence. It is the intention to reach beliefs, and to test those that are entertained, on the basis of observed fact, recognizing also that facts are without meaning save as they point to ideas. It is ... [this] attitude which recognizes that whole ideas are necessary to deal with facts yet they are working hypotheses to be tested by the consequences they produce. (1938b, 273).

The problems of inductive inquiry and the framing conditions are well illustrated by an example given by Royce in the case of trying to determine the color of marbles in an opaque bag. If we want to determine the color of marbles in a bag, we might draw one marble that is, perhaps, blue. Given a predetermination that the marble drawn will represent the other marbles in terms of color, we then conclude that the color of the marbles in the bag is blue. Rather than being a guess (the result of just picking a color without looking at the marble) or coming to a conclusion as a result of surveying all of the marbles (which would give us certainty), the inductive conclusion is based on a limited observation of the subject matter. Over time, we might be able to draw more marbles and, if we do, each successive marble will bring us closer to knowing what the color or colors of marble are in the bag.

---

[4] Peirce explains: "If in sampling any class, say M's, we first decide what the character P is for which we propose to sample that class, and also how many instances we propose to draw, our inference is really made before these latter are drawn, that the proposition of P's in the whole class is probably about the same as among the instances that are to be drawn, and the only thing we have to do is to draw them and observe the ratio. ... So that if the character P were not predesignate, ... we could not reason that if the M's did not generally possess the character P, it would not be likely that S's should possess that character" (2.737).

Royce argues that sampling procedures provide reliable guides if we suppose a relatively stable ontology partly independent of its being known (affirming, in effect, aspects of both realism and idealism). Royce observes "[in] general, if we choose a partial collection from a larger collection, and judge the constitution of the whole collection from that of the parts chosen, fixing our attention upon definable characters present or absent, in the partial collections, we are aided towards probable inferences by the fact that there are *more* possible 'samples' or partial collections, that at least approximately *agree* in their constitution with the constitution of the whole, then there are samples that widely disagree" (Royce, 1914, 329). In other words, the relation of a part to a whole is such that sampling a part can, in fact, tell us more or less reliably about the whole.

Consider a simple example. Suppose that we are presented with a bag containing four wooden blocks labeled *a*, *b*, *c*, and *d*, each of which is either painted red or white. Now suppose that you are asked to determine how many blocks are red and how many white by the following procedure: draw two blocks from the bag. If the blocks are both red, assert that all of the blocks are red. If the blocks are both white, assert that the blocks are all white. If one is red and one white, then assert that half of the blocks are red and half white. Let blocks *a* and *b* be red and blocks *c* and *d* be white. Consider the range of possible samples and what they will show. There are six possible pairs of blocks:

$$(a, b) (a, c) (a, d) (b, c) (b, d) (c, d)$$

Of these, only one pair, (a, b), if drawn, will have two red blocks and only one, (c, d), will have two white blocks. If either of these pairs is drawn, by the procedure you are following, you will give the wrong answer. However, four possible samples will give you the correct answer by the procedure above and so, if the samples are drawn at random, the structure of the entire collection is such that you have a four in six chance of giving the right result. Of course, the more samples taken, the more certain the result. Note that this example also illustrates the intersection of what the blocks "bring" to the experiment, decisions we make about what to see as significant, and a selection procedure of a particular character.

The example helps to illustrate a notion of *inductive validity*. Just as Habermas proposed notions of validity for intention and speech acts, the process of induction requires that certain procedures maximize the reliability of the process. If, instead of the procedure proposed by Royce, we adopted a procedure in which we extract a single block and let its color define the rest, we would have a smaller chance of choosing the right colors. Worse, if we have a procedure where we select a block and then guess a color other than the color of the block we chose, our chances of getting the right color decline still further. Inductive validity is relative to the determination of terms (blocks of certain colors), the horizon or boundary of the situation (the bag, on a given day at a given time), and how samples are selected and used.

This process of sampling models the more general process of induction in which a hypothesis that is meant to help order an indeterminate situation is tested. This form of induction, *generalization*, involves the selection of a sample of the subject class and, on the strength of the sample, infers a conclusion about the entire subject class. If the indeterminate situation is to ask the color of marbles in a bag, one way of answering the question is to draw a sample from the bag. The hypothesis is "if some percentage of the sample is a certain color, then that percentage of the whole is that color." The sampling process then establishes an answer to the original question and establishes a guide for future expectations and action that will hold unless and until some experience requires a new hypothesis.

Of course, the accuracy of the conclusion will depend upon the character of the sample selected. As Royce observes, the ability of a sample to lead to a successful induction will depend upon two factors: "the number of instances that have been empirically observed in [the process]," and "upon the fairness with which these facts have been chosen" (1914, 321). Following Royce, the character of the sample selected (and so the relative success of the argument) can be evaluated in terms of three standards.

1. *The sample should be random*, where a random sample is one in which each member of the class or collection under consideration has *an equal chance of being selected* relative to the characteristic under consideration. Suppose that we are interested in determining how many students at a university are concerned about racism on campus. Now, suppose that the class in question is all students at the university and that our sample is taken entirely from people entering and leaving the office of the Black Student Union. The selection procedure (which includes, perhaps, a higher percentage of students of color than might enter other offices or buildings on campus) is not structured in a way that gives all students an equal chance of being selected, just as it would not if the sample were taken exclusively at the door of the math library or a particular residence hall or at a volleyball game. By sampling at the entrance of the Black Student Union, it is likely that we will get a result that does not reflect the racial/ethnic character of the student body. Instead, we might select a sample by choosing every 20th student on a roster including all students sorted by last name. Although the method would not be perfect (it may include people who have just left the university and not include people that have just enrolled, for example), it would nevertheless be much closer to a truly random sample because all (or nearly all) students will be available for selection by a method that is independent of the characteristic under consideration.

2. *The sample should be appropriately related to the conclusion*, which depends, in turn, on the size and relative uniformity of the class considered and a determination of an "acceptable degree of error." Imagine a bag containing 50 red and 50 blue marbles from which we choose a single

marble. The result will tell us only that the marbles are all red or blue depending upon which we draw. If, on the other hand, we randomly choose two marbles, there is an even chance that one will be red and one blue. In this case, since the population of the class has only two colors and they are evenly distributed, a sample of two marbles would be better than one. If the population had three or four colors, however, a sample of two cannot give sufficient information to represent all of the colors present.

When samples are selected, they represent a portion of the total population included within the horizon of the situation. Even a random sample, when properly done, may not adequately represent the whole sampled. The difference between the percentage of the sample with a certain characteristic and the percentage of the whole population with that characteristic is called the *sampling error*. If one chooses a single marble from a bag of blue and red marbles where half of the marbles are blue, then while 100% of the sample is blue, only 50% of the population is blue. The sampling error is 50%. If 20% of the marbles in the bag are blue, then the sampling error is 80%. If, in a survey of Americans, the sample reports that 35% are in favor of invading oil rich countries, while, in fact, only 30% are in favor, the sampling error is 5%. Of course, in very large populations, it may be practically impossible to determine the "actual" number of whose who hold the view. While it is possible to correct for errors by taking larger samples and ensuring that the samples are randomly chosen, so long as the sample is fewer than the whole, there is a possible sampling error in the result.

3. Finally, *samples must be "objective"* in the sense that their selection is not based on any "psychological" factors. In the case above, our sample of university students would be poor if we only selected among the students going in and out of the Black Student Union office. Arguably, this poor sample would not be a product of particular psychological factors (e.g., race prejudice) since we could presumably have chosen the spot randomly and applied a mechanical selection process (e.g., we could have selected every third person walking through the door). In this case, the sample would still be problematic because it does not make all students available for selection. If the location had been chosen because we want to have a better chance of selecting minority students (and so reporting a higher percentage of minority students) then psychological factors have come into play. When the sample is not selected by some mechanical means but, for example, according to people we like or people we do not like or in light of the results we desire, then the sample is flawed. In this case, a sample based on one's race prejudices that led to the selection of only white students or only black students would generate a poor sample.

5.3   ANALOGICAL ARGUMENTS

While some ampliative arguments generalize from parts (samples) to larger wholes, others are comparative in character. *Analogical arguments* extend relations found in one thing, event or category to another that is similar in relevant ways. One sort of analogy compares two classes by asserting their similarity and that what follows from one class follows from the other. One might argue, for example, that since tulips are like daffodils, and daffodils need full sun to flourish, then tulips also need full sun to flourish. In this case, the argument has the rough form of

> All tulips are like daffodils (in some relevant ways).
> All daffodils are plants that need full sun.
> So, all tulips are plants that need full sun.

While we usually organize analogies in this way, if the premises are reversed so that "All tulips are like daffodils" is put in its proper place as the minor premise, the argument form is related to the first figure, mood AAA. Of course, the copula in the minor premise is not in standard form because it doesn't say that tulips *are* daffodils, but rather that they *are like* daffodils. This copula claims that the relation between the categories is suitable to warrant the conclusion. The standards for this sort of comparison become the principles for the validity of analogical arguments.

Another version of analogical argument places things in relation in a way that resembles inductive generalization:

> Labor leaders advocate for special interests, depend on support from their constituency, are accountable to their supporters, and tend not to be very independent in their thinking.
>
> Politicians advocate for special interests, depend on support from their constituency, and are accountable to their supporters.
>
> So it is likely that politicians tend not to be very independent in their thinking.

The argument relates the characteristics of labor leaders to those of politicians in a way that extends the characteristics attributed to politicians beyond the "facts" at hand. As in the first case, the limited premises lead to an amplified conclusion and, in each case, things that are not in the same category are used to ground the amplification.

In each case, the conditions governing inductive validity apply. In order that the terms of the analogy lead to the conclusion proposed, they must be determinate. It must be the case, for example, that daffodils and tulips can be identified for purposes of the analogical comparison in a context that is bounded in appropriate ways. If this situation did not restrict the samples to a particular stage in the annual growth cycle of the plants, the analogy might not yield the desired conclusion. For example, if one compared tulips in full bloom with dormant daffodils, the analogy about needing full sun would not hold. The sampling condition holds as well. In the second case, for example, the characteristics of labor leaders and politicians are identified in a way that seems relevant to the argument at hand and not obviously influenced by an overt interest in lauding (or damning) either group.

Analogical arguments fit the general pattern of induction because, like generalizations, they help to explain and order indeterminate situations. Rather than sampling, analogy is a process of proposing one thing or event in which a certain relation holds that is similar in appropriate ways to the thing to be explained. The hypothesis is that, for example, "if b follows from a in c, then b follows from a in d, where c is similar to d." The hypothesis in this case is a conditional relation between two three-term relations ("x follows y in z"). As in every other kind of induction, the process amplifies what we know (a relation in the context of c) to something that requires explanation (the meaning of d), and in finding an appropriate analogy, leads to a particular answer to the initial problem. As in the case of generalization, analogical arguments provide a means resolving indeterminate situations, but remain subject to revision and rejection as we take action and the expectations set are met or not met.

The general conditions for inductive validity can help inform a set of more specific standards that play a role in finding analogical arguments valid or invalid. These standards include adequacy of the sample, relevance of the comparison, sufficient similarity between the things compared, the relative unimportance of differences between the things compared, and the strength of the conclusion relative to the comparison offered. Each standard can be used as a critical tool for examining analogical arguments.

1. The analogy must be based on an *adequate sample.* To argue that tulips will not grow on the north side of a house because I once planted a daffodil on the north side and it died may not present an adequate ground for the comparison. If the connection between tulips and daffodils is to hold, we want to have some larger set of samples in order to make the point.

2. Analogies can be evaluated in terms of the *relevance of the comparison.* Imagine that Eloise is applying for a job as a lawyer in a large firm. After her interview, the partners in the law firm get together to evaluate Eloise for the position. One partner observes that he once knew a woman named Eloise and she was not reliable and so, he concludes, this Eloise may also be unreliable. A second partner replies that she has known a number of graduates from Eloise's law school. They have all seemed well prepared, effective in the courtroom, and good colleagues in the office. The partners might evaluate the two arguments in terms of the relevance of the comparisons to the conclusions that are offered. In the first case, while reliability might be a relevant characteristic in terms of which to make the decision, the comparison offered does not seem relevant to the conclusion. In Western societies, that two people have the

same name usually means very little about whether they have anything else in common. The second argument, however, both involves characteristics that seem relevant to the decision and they are offered in terms of a comparison that also seems relevant. In light of the arguments, one hopes that Eloise got the job.

3. While the standard of relevance (both of characteristics selected for comparison and of the things compared) provides a common ground on which to base a conclusion, it is also important that the things compared also are *sufficiently similar* in relation to the conclusion. Consider Brad, a student planning his work for his first course in math. He had heard from a friend that several students passed the course the previous year by attending every class. Brad resolved to attend every class in order to pass the class. While it is certainly possible that attending class might be required to pass, it is not necessarily the case that it is enough to pass. In this case, the analogy between Brad and the other students who passed the class does not appear to be sufficient to warrant Brad's conclusion.

4. At the same time, even if the similarities are significant, the differences between the things or events compared may undermine the conclusion. A valid argument will be one where the *differences between the things compared are unimportant* to the argument. A college adviser spoke with Amber about her college plans. The adviser explained that Amber and another student, Jill, were very similar. Both Amber and Jill had A averages and good test scores. Both had a college preparatory curriculum and both were good writers. Since Jill got into a highly selective university, Amber would too. What the adviser did not mention is that Jill's parents both graduated from the university she attends. Further, Jill was a state ranked volleyball player in her senior year of high school and was the president of the student council. Amber's parents attended a state college with open admissions and Amber is involved in very few extracurricular activities because she works 15 hours a week. Is the adviser's argument inductively valid? In some cases, even if the similarities are relevant and sufficient in some ways to warrant the conclusion, if the differences in the things (or students) compared are significant, the conclusion may not follow.

5. Finally, in cases where the sample is adequate, the comparisons are relevant, the similarities sufficient, and the differences unimportant, there remains the need to make sure that *the strength of the conclusion* is appropriate to the evidence offered. In the case of Amber's application to college, the adviser might have concluded that there was a good chance that she would get in if the factors that distinguished her from Jill did not play a large role in the decision. Consider another case. After getting her job at the law firm, Eloise buys a new hybrid car. Her friend, Paula, already had one of the same make and model, and was routinely getting 55 miles per gallon. When Eloise bought her new car, she con-

cluded by analogy that she would also get 55 miles per gallon. Unless Paula and Eloise are exactly the same kind of driver (and driving in the same conditions with the same maintenance and so on), Eloise's conclusion may be too strong. Had she concluded that she would get around 55 miles per gallon, her conclusion would be a better fit with the analogy she used.

Since all of these standards are imprecise, they lead to the question of whether or not they amount to standards at all or, if they are, whether they are sufficient standards for arguments to be valid in any significant sense. In response to this concern, one should recall the discussions of validity in Chapters Two and Three. In the end, validity (even formal validity) depends upon the acceptance of and familiarity with certain expectations about how arguments work. These expectations involve habits of interaction, purposes, and cultural and institutional constraints and affordances. What marks the end of an inquiry or the success of an argument is not simply the satisfaction of formal standards, but a felt change in the situation.

In broader philosophical terms, these changes may be understood as an aesthetic aspect of inquiry and argument. In the way that poets recognize the word or phrase they have been seeking to finish a line, or the way a musician finds the "right" note or line in an improvisation, participants in inquiry and argument recognize that a comparison is sufficient or that the differences at hand are unimportant relative to the issue. They recognize when a sample is adequate and when a conclusion is too strong. When there is disagreement among participants, they have the option of discussing the disagreement and arriving collectively at a recognition that the analogy in question works or fails.

Is this validity? In an important sense it is, since when the argument includes determinate terms and a bounded context, and if the sample is fair, the conclusion will follow with a degree of certainty. The difficulty of arriving at an argument that satisfies the conditions of inductive validity is vastly more difficult than identifying formally valid arguments, but it is not different in kind. In both cases there is an understanding that certain expectations must be met and that, if they are, all participants ought to accept the outcome.[5] If some do not, it might be because they are simply recalcitrant, or it might be that the operative expectations differ in some important way. In this case, the claim of validity does not lead to the rejection of the views of the one who disagrees, but rather suggests that further conversation is needed.

---

[5] That formal validity also requires a framework comparable to that of inductive validity, consider Graham Priest's work in his book *In Contradiction* (2006). Though Priest may not draw this conclusion, an argument can be made about the importance of the expectations of the participants in inquiry and argument in relation to situations that lead to diverse forms of formal validity.

5.4 EXERCISES

For each of the following examples of induction or analogical reasoning, decide whether or not the argument is valid in light of the standards discussed in 5.2 and 5.3. Explain your answer.

1. A random sample of 1,000 Oregonians found that 51% favored the war in Iraq. Therefore, 49% of all Oregonians are opposed to the war.

2. "Now if we survey the universe, so far as it falls under our knowledge, it bears a great resemblance to an animal or organized body, and seems actuated with a like principle of life and motion. A continual circulation of matter in it produces no disorder: a continual waste in every part is incessantly repaired; the closest sympathy is perceived throughout the entire system: and each part or member, in performing its proper offices, operated both to its own preservation and to that of the whole. The world, therefore, I infer, is an animal, and the Deity is the *soul* of the world actuating it, and actuated by it." (Hume, 1779, 82–3)

3. Among the students taking the history of philosophy this term, John, Shane, Michael, Roberta, Ofelia, Louise, Abigail, Suzanne, and Flora prefer the philosophy of Kant to that of Hegel. I have yet to talk with Jennifer and Martin, but I suspect that they will prefer Kant as well.

4. Most long-time Democrats are opposed to a long military involvement in Iraq. My friend Elmer is a long-time Democrat so my friend Elmer is someone who opposes a long military involvement in Iraq.

5. To allow banks and the financial sector to operate with little or no regulation is like letting a child operate in a candy store with little or no regulation. What we find is that they both will overindulge in whatever seems most immediately pleasurable. The child will devour any treat that promises even a hint of instant satisfaction and the financers will speculate on the riskiest of ventures if they show even the slightest promise of satisfactory returns. In the end, both will crash miserably from their self-induced highs, the child's coming from consuming too much sugar and the speculator's from holding too much bad debt.

6. A survey of San Diegans was taken outside of the football stadium directly after a Charger victory. The survey sampled over 5,000 people and was restricted to only San Diegans of voting age. That survey revealed that 75% of San Diegans support the public financing of a new stadium for the Chargers to play in. This survey therefore seems to show that the stadium measure will pass easily.

7. A very reliable study found that at least 80% of the human population lives on less than $10 a day. In my current logic class there are about 100 people registered and needless to say they all belong to the human popula-

tion. It would therefore seem to follow from this study that roughly 80 people in my logic class live on less than $10 a day.

8. Most waiters and waitresses that work in downtown, when tips are factored in, make over $30 an hour. None of my good friends make over $30 an hour, even when tips are factored in. Therefore it is safe to conclude that none of my good friends are waiters or waitresses that work in downtown.

9. "I will tell you, I [Socrates] replied; justice, which is the subject of our inquiry, is, as you know, sometimes spoken of as the virtue of an individual, and sometimes as the virtue of a State.

   True, he [Adeimantus] replied.

   And is not a State larger than an individual?

   It is.

   Then in the larger the quantity of justice is likely to be larger and more easily discernible. I propose therefore that we inquire into the nature of justice and injustice, first as they appear in the State, and secondly in the individual, proceeding from the greater to the lesser and comparing them.

   That, he said, is an excellent proposal.

   And if we imagine the State in process of creation, we shall see the justice and injustice of the State in process of creation also.

   I dare say." (Plato, 1892, 48–9)

10. A survey was conducted of women's attitudes toward education. A sample was selected randomly from among a group of Minnesota women who had dropped out of high school. Based on this study, funded by the "Protect Traditional Values Action Committee," we have determined that among Minnesota women, 82% feel that education had little or no effect on their quality of life.

11. I heard from a pretty reliable source that in 2007 the average American household income rose 1.3% to $50,233. My sister Lorena lives in a pretty average American household. Therefore, her income in 2007 must have risen 1.3% to $50,233.

12. "If you cut up a large diamond into little bits, it will entirely lose the value it had as a whole; and an army divided up into small bodies of soldiers, loses all its strength. So a great intellect sinks to the level of an ordinary one, as soon as it is interrupted and disturbed, its attention distracted and drawn off from the matter in hand: for its superiority depends upon its power of concentration—of bringing all its strength to bear upon one theme, in the same way as a concave mirror collects into one point all the rays of light that strike upon it." (Schopenhauer, 1908, 127–8)

5.5   CAUSAL ARGUMENTS

In 1843, John Stuart Mill (1806–1873) published *A System of Logic* as a response to the widely accepted philosophical idealism of the time. He concluded of the dominant philosophy:

> The notion that truths external to the mind may be known by intuition or consciousness, independently of observation and experience, is, I am persuaded, in these times, the great intellectual support of false doctrines and bad institutions. By aid of this theory, every inveterate belief is enabled to dispense with the obligation of justifying itself by reason, and is erected into its own all-sufficient voucher and justification. There never was such an instrument devised for consecrating deep-seated prejudices. (Mill, 1924, 157)

Mill's response was to offer a conception of logic and scientific method based on the process of learning from experience. His goal in *A System of Logic* was to develop a set of rules in terms of which induction could be understood and evaluated. "The task to be performed was that of generalizing the modes of investigating truth and estimating evidence by which so many important and recondite laws of nature have, in the various sciences, been aggregated to the stock of human knowledge" (1875, 4).

In order to explain how science has produced its "recondite laws," Mill proposed five canons of induction which, when presented in arguments, amount to five inductively valid argument forms. These canons describe the methods of agreement, difference, the joint method of agreement and difference, residue, and concomitant variation.

The methods Mill proposed were concerned with causation, but are based on direct comparison of cases and conditions that are actually shared or not shared. These methods accepted the idea that we can identify necessary and sufficient conditions of the occurrence of some thing or event by comparing cases in which different conditions are present. A *necessary condition* is one that is required if a certain effect is to follow. For example, it is necessary that students attend 8 of 10 discussion sections for a particular class in order to get a grade higher than a C–. Notice, however, that attending all of the sections will not necessarily earn a student a better grade than a C–. A *sufficient condition*, on the other hand, is "enough" for a certain effect to follow. So it is sufficient to get a failing grade on the mid-term exam by simply skipping the exam. One could, in fact, take the mid-term exam and still get a failing grade, but simply not taking it would be enough.

Mill's causal arguments, like generalizations and analogy, are inductive arguments in that they provide support for hypotheses that transform indeterminate situations into more determinate ones by proposing that the problem at hand is the effect of some particular cause. In this case, the hypothesis is that certain kinds of evidence constitute justification for the conclusion that a causes b, where b is taken as the thing or event we are trying to explain. If Charles falls ill and we want to know why, we seek to determine what caused Charles' illness. The hypothesis, something, a, caused his illness, b, can then be examined in order to find out how to cure Charles and, perhaps, prevent such illness in the future. Mill's methods provide a framework for determining causes and whether they are necessary or sufficient.

The first canon of induction, governing the *method of agreement*, provides a means of determining necessary conditions.

> If two or more instances of the phenomenon under investigation have only one circumstance in common, the circumstance in which alone all the instances agree, is the cause (of the effect) of the given phenomenon. (Mill, 1875, 214)

In short, the method of agreement looks for causes that are common in all cases in which the effect also occurs. Suppose eight students go to dinner together at Joe's Cafe. Later that night, each one becomes ill. The next day the eight people compare their activities over the last few days and determine that the only thing that they all did in common was to have dinner at Joe's. Since there are no cases in which someone became ill and did not eat at Joe's it appears that eating at Joe's was necessary for the illness to occur. Put another way, the method of agreement leads to the rule that something is *not* a necessary condition if it occurs without the effect occurring. It is also possible that one can get sick for reasons other than eating at Joe's, but the case at hand does not provide any evidence to that effect. Given the determinate terms of the problem and the boundaries set for the example, there are no cases of getting ill that did not follow eating at Joe's. Of course, the common factor must be in some way relevant to the effect in question. If all eight people found that they all walked upright during the course of the day, we would expect that they would look for a more relevant common activity.

The *method of difference* provides a means of determining sufficient conditions. According to the second canon

> If an instance in which the phenomenon under investigation occurs and an instance in which it does not occur, have every circumstance in common save one, that one occurring only in the former, the circumstance in which alone the two instances differ, is the effect of the cause, or an indispensable part of the cause, of the phenomenon. (1875, 215–6)

Suppose Albert and Mike share a dorm room and have the same classes. In fact, they are together most of the time. One morning Mike wakes up with a fever and a stomachache. They compare their activities of the previous day and discover that they did everything together except eat dinner. Mike, of course, ate at Joe's. In this case, the difference between Mike and Albert is eating at Joe's and so it appears that eating at Joe's was sufficient to get sick. The other activities they considered are ruled out as sufficient to cause the illness by the rule that something is *not* a sufficient cause if it occurs and the effect does not. The evidence, of course, does not guarantee that eating at Joe's will make you sick (one might eat at Joe's and pass on the coleslaw, for example), but it does give strong evidence that eating at Joe's can cause illness. There may be plenty of other causes of illness, some of which are necessary. In the case at hand, however, the evidence only supports the conclusion that eating at Joe's is sufficient to get sick.

The third canon of induction *combines agreement and difference* in order to identify causes that are both necessary and sufficient conditions.

If two or more instances in which the phenomenon occurs have only one circumstance in common, while two or more instances in which it does not occur have nothing in common save the absence of that circumstance, the circumstances in which alone the two sets of instances differ is the effect, or the cause, or an indispensable part of the cause, of the phenomenon. (1875, 221)

Consider again the visit to Joe's. In this case, Mike takes three friends, Natalie, Carlos, and Renee. Mike ordered a salad, a vegetarian burger, onion rings, and a shake. Natalie also ordered a vegetarian burger, but also had coleslaw, soup, and a latte. Carlos ordered a cheeseburger and coleslaw with onion rings and a soda. Renee had a cheeseburger and fries, a salad, a shake, and coffee. Later that day, Carlos and Natalie became ill while their friends did not. Using the method of agreement and the rule that a cause is not necessary if it occurs in the absence of the effect, we can eliminate salads, shakes, and coffee as necessary conditions. Using the method of difference and the rule that a cause is not sufficient if the effect occurs without it, we can rule out sodas, onion rings, fries, vegetarian burgers, cheeseburgers, lattes, and soup. The sole remaining candidate for a necessary and sufficient condition for getting ill is eating Joe's coleslaw.

The fourth canon of induction is based on the *method of residue*. This method identifies complex causes.

Subduct from any phenomenon such part as is known by previous inductions to be the effect of certain antecedents, and the residue of the phenomenon is the effect of the remaining antecedents. (1875, 223)

Suppose that a gardener wants to find out why her garden never produces vegetables. After consulting some neighbors, she decides that the problem might be a lack of water, a lack of sunlight or grazing deer. Over the next few weeks, the gardener makes sure to water the garden regularly and the plants begin to grow more energetically. She notices that several branches of a large tree block the sun to the garden for most of the morning. She trims the tree to let the sun reach the garden. She notices that the plants begin to produce beans, peas, and small tomatoes. After a week of good sunlight and water, she notices that most of the vegetables have vanished and concludes that deer from the nearby woods have been dining in her garden. In this case, given three possible causes, the first two are taken into account (by watering and trimming the tree) and they seem to address the problem, though not completely. The residue—the left over effect—is then attributed to the remaining cause.

Finally, consider a case in which on one day in a residence hall on campus there is a high incidence of stomach ailments and student absences from class. On other days, there are very few cases. On examination, it turns out that days when many residents of the hall eat at Joe's instead of the campus dining room are followed by days when many people are sick. We can conclude that since the incidence of eating at Joe's and of illness increase and decrease together, the illnesses are, in fact, caused by eating at Joe's. This argument is the type Mill called *concomitant variation*. The fifth canon of induction is

> Whatever phenomenon varies in any manner whenever another phenomenon varies in some particular manner is either a cause or an effect of that phenomenon, or is connected with it through some fact of causation. (1875, 227)

While eating at Joe's may be determined as a cause of getting sick because incidents of illness rise when more people eat there, like the other methods, relevance remains an important consideration. For example, one might find, in a survey of ice cream eating and death by drowning that the ice cream eating increases and decreases at nearly the same rate as drowning deaths. Concomitant variation would seem to warrant the conclusion that eating ice cream causes drowning. While those suspicious of the dairy industry might agree, the example is a better illustration that concomitant variation can also establish a simple correlation without a causal relation. Recognizing the relevance of the proposed causes to one another provides a means of rejecting conclusions that are only correlations.

5.6 EXERCISES

5.6.1. In each of the following decide whether the cause mentioned is a necessary condition, a sufficient condition, both or neither.

1. Turning 18 makes a U.S. citizen eligible to vote.

2. Providing the post office with your forwarding address causes all your mail to be sent to your new address.

3. Being an accomplice to burglary causes one to be convicted of a crime.

4. In order for racial segregation to be eliminated, Jim Crow laws needed to be done away with.

5. Carlos Rodriguez has U.S. citizenship because he was born in the state of New Mexico.

6. Understanding a philosopher's thought comes from carefully reading her work.

7. Between 1910 and 1931 various U.S. states (e.g., Tennessee, Louisiana, Texas, Arkansas, Mississippi, North Carolina, Virginia, Alabama, Georgia, and Oklahoma) adopted "one-drop" legal statutes with regards to racial classification. According to these legal statutes any African ancestor was cause to classify a person as Black.

8. Attacking a sovereign nation for non-self-defense reasons and without an authorization for military intervention from the United Nations Security Council will cause a nation to be in violation of international law.

5.6.2 Consider each of the following examples. Determine which of Mill's methods was used and explain your answer.

1. For the last few weeks I had been getting a terrible night's sleep and in the morning I found it almost impossible to get out of bed. The only things I noticed that had been different from before was that I started smoking again, I got a new stressful job, and I have been putting myself to sleep with cheesecake every night. I have since stopped smoking and I have cut out the cheesecake, but my sleep is still not great and I'm still finding it difficult to get up in the morning. I conclude that the new stressful job is responsible for my problems sleeping and my terrible time getting up in the morning.

2. Two couples that on paper appeared to be qualified candidates for being adoptive parents both had their adoption request rejected and wanted to figure out why. One of the couples in question was white, middle class, and lived in a downtown condo; the other was black, upper class, and lived in a gated suburban community. The differences between the two went on, but it happened that the two couples were both same-sex couples seeking to adopt a child in a conservative state. They therefore concluded that this last point was the cause of their rejected adoption request.

3. In 2008 senators Ted Stevens and Rodger Wicker ran for re-election. Both Stevens and Wicker have very similar views and were incumbent Republican senators in states that traditionally vote heavily Republican (Alaska and Mississippi, respectively). Rodger Wicker won re-election easily, but Ted Stevens lost narrowly to his unknown challenger Mark Begich. Stevens, however, was found guilty on seven counts of failing to properly report gifts a week before the election and Wicker was not. It is reasonable to conclude that Stevens lost because of his felony convictions (which were later overturned).

4. A variety of reports on development in Latin American and African nations have shown that despite the availability of valuable natural resources, the nations remain economically poor. These nations have little in common in terms of culture, history, and forms of government. However, it turns out that they all have received loans from the IMF and the World Bank. These loans come with strings attached that require the receiving nations to adopt certain internal economic policies that, one might conclude, account for the ongoing poverty of these nations.

5. In order to expand a highway, a very close-knit community was forced to relocate. The city knew that the residents of that community would retain hard feelings towards the city if concerns around financial compensation, decent new housing, continued access to good schools, and a sense of community were not met. The residents were all financially compensated fairly and relocated to decent housing where good schools were very accessible. Two years after the community was forced to relocate, however, there are still hard feelings from the former residents towards the city for its use of eminent domain. The cause of the hard feelings therefore seems to be that, in the relocation process, the sense of community could not be kept intact.

6. One state, where violent crime had become a very serious problem, decided to conduct a study to see if violent crime could be reduced with the implementation of harsher sentences. Two graphs were drawn up, one displaying the crime rate and the other the rate of harsher sentencing, and were super-

imposed on each other. They showed no historical correlation between a rise in harsher sentences and the rate of violent crime. When, however, the graph displaying the rate of violent crime was superimposed on a graph displaying the historical rate of poverty, one could see clearly that when there was an increase in the rate of poverty there followed closely an increase in the rate of violent crime; when the poverty rate went down, it was followed closely by a drop in the rate of violent crime. The conclusion was that to lower violent crime harsher sentences were not needed, but in fact the problem of poverty needed to be dealt with first.

7. In the county of San Diego, California, there are two high schools (i.e., La Jolla High and San Ysidro High) in which the student body is relatively similar (e.g., in terms of race, gender, and standardized test scores), the curriculum is exactly the same, the classes are the same, the graduation requirements are the same, and the teachers and administrators are also similarly educated and equally caring about their students' well-being and education. We find, however, that the students from La Jolla High graduate at a higher rate and are more likely to go to a four-year university than are the students from San Ysidro High. The only difference is that there is a great disparity in the wealth of families of the students and there is a radical disparity in the real-estate value surrounding each of the schools. At Torrey Pines High School, which has nothing like the student body of either San Ysidro or La Jolla, and also has very different curriculum, graduation requirements, teachers and administrators, the graduation rate and the likelihood of its students going to a four-year college is similar to that of La Jolla High. What Torrey Pines and La Jolla share is that they both have a similar proportion of wealthy students and the surrounding real estate values are very high.

8. Average annual surface temperatures of the Earth have increased significantly since the 19th century and ten of the warmest years on record have occurred since 1995. It is also the case that an increase of carbon dioxide in the atmosphere increases average temperatures and a decrease of carbon dioxide comes with a decrease in temperatures. Humans are a major source of carbon dioxide production, especially through manufacturing and the use of fossil fuels. These two forms of human activities have increased significantly since the 19th century. We conclude that human activity is a key cause of global warming.

9. "Though the encouragement of exportation and the discouragement of importation are the two great engines by which the mercantile system proposes to enrich every country, yet with regard to some particular commodities it seems to follow an opposite plan: to discourage exportation and to encourage importation. Its ultimate object, however, it pretends, is always the same, to enrich the country by the advantageous balance of trade. It discourages the exportation of the materials of manufacture, and of the instruments of trade, in order to give our own workmen an advantage, and to enable them to undersell those of other nations in all foreign markets; and by restraining, in this manner, the exportation of a few commodities of no great price, it proposes to occasion a much greater and more valuable exportation of others. It encourages the importation of the materials of

manufacture in order that our own people may be enabled to work them up more cheaply, and thereby prevent a greater and more valuable importation of the manufactured commodities." (Adam Smith, 1817)

5.7 PROBABILITY

While generalizations of the sort discussed in 5.2 are attempts to infer the features of a larger group from a sample, *probabilistic arguments* attempt to determine how patterns can be more precisely *extended to new cases*. In the case of Sarah's tardiness, for example, we attempted (without much precision) to infer whether or not Sarah would be late for an appointment. Probabilistic arguments add precision to the basic form of inductive generalization. In order to extend a claim to a new case, one must have an initial sense of the likelihood of the claim applying to a new case. This initial determination of likelihood, which we will call *initial probability*, sets the stage for our calculations about the application of a claim. In the case of Sarah, we began with the idea that she is usually late. Though this is not a precise measure, it nevertheless can be used as an initial probability from which to calculate whether or not Sarah will be late this time. When initial probabilities are more precisely established, then we can use a formal deductive process to set expectations about future events or to assess whether or not a claim will apply to a new case.

Initial probabilities, however, are not simple givens, but rather are "takens" based on different ways of understanding what has gone before. There are three such approaches or interpretations: classical theory, subjective theory, and the theory of relative frequency. In each interpretation, the initial probability can express the probable occurrence of some event (e.g., Sarah's on-time arrival), the presence of some property or characteristic (e.g., whether a thing is a tiger or a leopard), or the truth or falsity of an assertion (e.g., the assertion "It will rain tomorrow."). Given the general approach to meaning taken here, differences raised by these different interpretations of initial probability can be set aside in order to focus on how they set the stage for probabilistic arguments.

The *classical theory* for determining initial probability holds that, given a fixed set of possible outcomes, the probability of a particular event, E, is the number of "favorable outcomes" (cases in which E can occur) divided by the total number of outcomes. In the throw of a "fair" die, for example, the probability of rolling a six is the number of favorable outcomes (1 in this case since there is only one six on a die) divided by the total number of outcomes (6 since it is a six-sided die), or 1/6 or "1 out of 6." The classical interpretation depends upon two assumptions: that all outcomes are equally possible (known as the *principle of indifference*), and that all possible outcomes are included in the calculation. The classical theory works fairly well for situations that involve limited outcomes and independent events (where one outcome does not effect other outcomes). However, when outcomes are not indifferent and when the set of outcomes is large or infinite, another approach is often taken.

The *subjective theory* understands probability, not in terms of actual outcomes, but in terms of "confidence" that a particular event will occur. In this case, the probability that Sarah will be late, for example, is given in terms of the odds that she will arrive late. Since she is often late, I might give 5 to 1 odds that she will be late today. The probability of her late arrival, therefore, will be the number of "favorable" outcomes (i.e., the number of outcomes where she will in fact be late, 5) divided by the total number of outcomes (i.e., five late outcomes plus one on time arrival for a total of 6). In this case, then, the subjective probability of Sarah's late arrival is 5/6. The assignment of odds is based on the odds-maker's "sense of" or "feeling about" the case at hand. While determining odds might involve taking stock of past events and current circumstances, odds are not "objective." Different odds-makers can give different odds for the same event. We might give 5 to 1 odds on Sarah's being late while someone else might give 10 to 1 odds. Odds are, of course, expressed numerically and so can serve as the starting point for a calculation of probability and, when odds-makers differ about the chances of a given event, their diverse odds can be averaged so that the result is an expression of a collective sense of the event in question.

The third interpretation of probability is called *relative frequency theory*. In this case, the probability of a particular event, E, is calculated in light of what can be observed. Here, the probability of E equals *the number of observed favorable outcomes* divided by *the number of observed outcomes*. Since none of us have observed every case of Sarah's arriving somewhere, we can calculate the probability of her arriving late today by dividing the number of times she has arrived late for a meeting with us (say 12 times) by the number of times she has arrived for a meeting with us (on time and late, say 16 times), or 12/16 or 3/4. In effect, given the relative frequency of the pattern of arriving late, there is a 75% chance that she will arrive late this time.

The rules for calculating initial probability are fairly simple to formulate. Consider the following:

> *Classical Theory:* $P(p) = f/n$, or "the probability of p is the number of favorable outcomes, f, divided by the total number of possible outcomes, n." The probability of drawing a red marble from a bag filled with 10 red marbles and 30 white marbles is the number of favorable outcomes (10) divided by the total number of possible outcomes (40) or 10/40 or 1/4 or 25%.

> *Subjectivist Theory:* $P(p) = x/(x + y)$. Given the odds of a favorable outcome, x-to-y, (e.g., 4 to 1, 5 to 1), the probability of p is x divided by $x + y$. If the odds are 4 to 1, then $P(p) = 4/(4 + 1) = 4/5$. Odds are determined "subjectively," that is, by whatever means chosen by the person giving the odds (this can be a guess or the result of some more or less systematic calculation of past results). Given the odds selected, probability is then calculated.

Relative Frequency: $P(p) = f_o/o_o$, or "the probability of p is the number of favorable outcomes observed, f_o, divided by the number of outcomes observed, o_o." If you observe the women's basketball team winning 8 of the 12 games you attended, then the probability of a win is the number of favorable outcomes you have seen (8) divided by the number of outcomes you have seen (12) or 8/12 or 2/3 or about 66%.

Once the initial probability for an event or the truth of a claim has been determined by whatever means, one can then use a *probability calculus* to extend the initial probability to further cases. This process, like all induction, is constrained by the general principles of induction proposed in 5.2. The calculation of initial probability serves to determine the terms under consideration and the particular interpretation used sets the parameters for what will count as a fair sample. The "calculus" then provides the boundaries for the situation. Consider any term of a situation, call it A. If A is a term (an assertion, another kind of action, an event), then its probability can be expressed

$$P(A) \geq 0 \qquad (1)$$

that is, the probability of A's being true (or of A's occurrence) is greater than or equal to 0. The definition claims that any particular term of the situation that is not necessary either will not occur and so has a probability of 0 or will occur and so will have a probability greater than 0. The probability of rolling a six with a fair die is, as we observed, 1/6 or .17 and so P(rolling a six) = .17 (and is greater than or equal to 0).

While (1) determines the probabilistic character of individual terms, the boundaries of the situation must be established as well. Formally, the boundaries of the situation can be called the *universe of discourse* and represented by 'U'. The universe can then be defined

$$P(U) = 1 \qquad (2)$$

that is, the sum of all the probabilities of all the events (or terms) of the situation is 1 (or 100%).

Initial probabilities of individual terms can be represented as follows

$$P(x) = n \text{ such that } 0 \leq n \leq 1 \qquad (3)$$

Initial probabilities can be given for events that are conditionally related so that x occurs given that y occurs so that

$$P(x, y) = n \text{ such that } 0 \leq n \leq 1. \qquad (4)$$

"P(x, y)" is read "the probability of x given y."

Now suppose the initial probability of A is .5 or 50%. Given that the total probability of the universe is 1 and the probability of A is .5, then the probability of A not happening (written as '¬A') is also .5. If the probably of rain on a given day is 40%, then the probability of having no rain is 60%. This relation between A and ¬A (rain and no rain) is called *exclusive*.

Consider a case in which Max can choose between four activities: going to a study group, volunteering at a food shelf, going to a movie, or staying home. Let these alternatives be labeled A, B, C, and D, respectively. If this is the full range of possibilities for Max, then the sum of their probabilities will be 1 given the universe of discourse. This relation among terms is called *exhaustive*. It is easy to see that the relation between A and ¬A is exclusive and exhaustive while the relation between A and B is exclusive but not exhaustive.

Let '·' represent the two-place relation "and" (as in "A *and* B"), called a *conjunction*, and 'v' represent the two-place relation "A or B or both" called a *disjunction*.[6] The exclusive relation between terms can be represented

$$P(A \cdot \neg A) = 0 \tag{5}$$
$$P(A \vee \neg A) = 1 \tag{6}$$

Given that A and ¬A exhaust the possibilities, their probabilities must add up to 1 so that following should hold as well:

$$P(A) + P(\neg A) = 1 \tag{7}$$
$$P(\neg A) = 1 - P(A) \tag{8}$$

where (7) reads "The probability of A plus the probability of not-A equals 1" (or "The sum of the probability of A and the probability of not-A is 1") and (8) reads "The probability of not-A is 1 minus the probability of A."

Suppose the initial probability of Max going to the movies is 20% and the initial probability of his staying home is 20%. Since these two alternatives are mutually exclusive (Max cannot do both), then the probability of his doing both is 0 and of doing one or the other is the sum of the probability of going to the movies and staying home.

$$P(C \cdot D) = 0$$
$$P(C \vee D) = P(C) + P(D) \tag{9}$$

Now suppose that Max decides that he may want to have popcorn (call this 'E') which, it turns out, he might have whether or not he goes to the movies. His probability of having popcorn, whether or not he goes to the movies is .40. What is the probability that Max will go to the movies or have popcorn or both where his chance of going to the movies is .20? To see how this works, represent the universe as a Venn diagram where each circle contains one 'X' for each 10% chance of the activity happening (so there are two Xs for the movies and four Xs for popcorn). Since it is possible to go to the movie without getting popcorn, place one X in C and one X in the area where

---

[6] Disjunctions have two forms: inclusive and exclusive. An inclusive disjunction is true when either of the disjuncts is true or both are true. The exclusive disjunction, however, is true when one disjunct is true and one is false.

C and E overlap. Similarly, since Max may have popcorn whether or not he goes to the movies, place an X in the overlapping area and three more Xs in E. The total number of possibilities then is 6 (1 for going to the movies, 2 for going to the movies and having popcorn, and 3 for having popcorn). If we calculate the probability of C and E or both by adding all the chances together, the result would be $P(C) + P(E) + P(C \cdot E)$ and so $2/6 + 4/6 + 2/6$ or $8/6$. Notice that this leads to a probability that is greater than 1 and so impossible in the universe of discourse. The mistake is to double-count the overlapping possibilities. Instead, C or E or both can be calculated this way

$$P(C \vee E) = [P(C) + P(E)] - P(C \cdot E) \tag{10}$$

Note that I have added square brackets to indicate which terms relate to the '+' and '−' operators.

What if we want to know the probability of two events both occurring? Suppose A and B are two activities that are independent of each other and that the $P(A) = .20$ and $P(B) = .50$. The probability of both occurring is

$$P(A \cdot B) = P(A) \times P(B) \tag{11}$$

$$P(A \cdot B) = .20 \times .50 = .10 \text{ or } 1/5 \times 1/2 = 1/10 = 10\%$$

However, consider this example: A store manager observes that the probability of a male customer entering her shop on a Thursday morning (M) is 0.4. The probability of making a sale on a Thursday morning (S) is 0.6. The manager wants to know the probability of a male customer coming in the shop *and* making a sale on a Thursday morning, that is, the value of $P(M \cdot S)$. In order to determine this, we need to connect the events. Since we know that there is a chance that a sale will be made if a customer comes into the shop, then the chances of making a sale if a male customer comes into the shop is 0.6. This can be represented as $P(S, M) = .60$, that is, the probability of making a sale (S) given that a male customer comes into the shop (M). So the probability of both a male customer coming into the shop and a making a sale is .24.

$$P(M \cdot S) = P(M) \times P(S, M) \tag{12}$$

$$P(M \cdot S) = .40 \times .60 = .24 = 2/5 \times 6/10 = 12/50 = 6/25$$

While conditional probability has been offered as something to be determined initially, it can also be definied in light of the general conjunction rule just given. If general conjunction is defined as $P(x \cdot y) = P(y) \times P(x, y)$, then both sides can be divided by $P(y)$ so that conditional probability can be defined

$$P(x \cdot y)/P(y) = [P(x, y) \times P(y)]/P(y) = P(x, y)$$

and so

$$P(x, y) = P(y \cdot x)/P(y) \tag{13}$$

Suppose that I toss a fair die and it lands with an even number of dots facing up. What is the probability that I rolled a four? Let $P(E)$ be the probability that I rolled an even number, so

$$P(E) = 1/2$$

If $P(F)$ is the probability that I rolled a four, then $P(E \cdot F)$ is the probability of rolling an even number that is also a four.

$$P(E \cdot F) = 1/6$$

The probability that I rolled a four given that I rolled an even is

$$P(F, E) = P(E \cdot F)/P(E) = 1/6 \text{ divided by } 1/2 = 1/3 \text{ } or$$

$$P(F, E) = P(E \cdot F)/P(E) = .167/.5 = .33$$

So the conditional probability of rolling a four given that I rolled an even number is 33%.

Some situations can be described as a summary of terms that are mutually exclusive and exhaustive. The simplest such situation is one that involves only an element and its complement. In this case, the situation can be described by the "or" relation: A v ¬A. Imagine a more complicated situation. With each pitch at a baseball game, an umpire faces a choice: the pitch is a strike, a ball, a foul, a hit, or an out. Formally, this situation can be described as "S v B v F v H v O." Taken together, the probability of all of these possibilities will total 1. More generally, the situation can be described with a disjunctive proposition with a finite number of terms: E_1 v E_2 v ... v E_n. If we want to express the probability of some event A and each of the possible outcomes, then

$$P(A) = P(E_1 \cdot A) + P(E_2 \cdot A) + ... + P(E_n \cdot A) \qquad (14)$$

Suppose we would like to know the probability of a batter being out after the next pitch. In the simplest case, the batter has two strikes and no balls. There are two possibilities: either it is a strike or not a strike. If it is a strike, then the probability is 1 that the batter will be out. If it is not a strike, the batter still has (for example) a 15% chance of being out (since it could be a fly ball or a ground out). Given a projection of the pitcher's ability and the ability of the defense, figure 5.1 illustrates the relation, where O is an out and S is a strike:

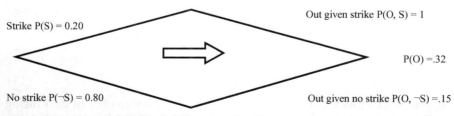

Figure 5.1   Graph of the Probability of an "out"

The diamond provides two paths to the calculation of an out. The top line projects a strike and, since there are two strikes, a 100% chance of an out. The bottom line supposes no strike and calculates the chance of a fly ball or play turning into an out. The probability on the right is the probability of an out on the given pitch in the situation as described. Notice that the equation a (below) is the *Principle of Total Probability*. Line b substitutes the conjunctions with their equivalents as given in the Principle of Conjunction (13).

a. $P(O) = P(O \cdot S) + P(O \cdot \neg S)$
b. $ = [P(S) \times P(O, S)] + [P(\neg S) \times P(O, \neg S)]$
c. $ = [.20 \times 1] + [.80 \times .15]$
d. $ = .2 + .12$
e. $ = .32$

The probability of the next pitch being an out is 32%.

Bayes' Theorem provides a way to calculate conditional probabilities in complex circumstances. Suppose that $P(H, E)$ represents the probability of the truth of hypothesis H given evidence E. Given the conditional rule,

$$P(H, E) = P(E \cdot H)/P(E)$$

Given the Principle of Total Probability, we can replace the denominator of the formula above so that

$$P(H, E) = P(E \cdot H)/P(E \cdot H) + P(E \cdot \neg H)$$

and, using the Principle of Conjunction,

$$P(H, E) = P(E \cdot H)/[P(H) \times P(E, H)] + [P(\neg H) \times P(E, \neg H)].$$

Finally, we can replace $P(E \cdot H)$ with $P(H) \times P(E, H)$ so that the relation between the hypothesis given the evidence for it can be represented by Bayes' Theorem (here the "Limited" version since it considers only two cases, H and $\neg H$):

$$P(H, E) = \frac{P(H) \times P(E, H)}{[P(H) \times P(E, H)] + [P(\neg H) \times P(E, \neg H)]} \tag{15}$$

The expanded version of the theorem considers any finite set of possibilities so that the general form can be expressed as follows, where $x_1 \ldots x_n$ represents the range of possibilities:

$$P(x_1, y) = \frac{P(x_1) \times P(y, x_1)}{[P(x_1) \times P(y, x_1)] + [P(x_2) \times P(y, x_2)] \ldots + [P(x_n) \times P(y, x_n)]} \tag{16}$$

The Theorem can be expanded to accommodate additional cases by simply adding further mutually exclusive and exhaustive cases to the denominator. Suppose that we are considering three hypotheses, one of which must hold. In this case, H_n, will represent each of the cases and the resulting theorem is

$$P(H_1, E) = \frac{P(H_1) \times P(E, H_1)}{[P(H_1) \times P(E, H_1)] + [P(H_2) \times P(E, H_2)] + [P(H_3) \times P(E, H_3)]}$$

To illustrate Bayes' Theorem consider five bags of marbles, one belongs to Amanda and four belong to Raphael. The bags are identical on the outside, but Amanda's bag contains eight red marbles and two white marbles and Raphael's each contain three red marbles and seven white. Now suppose that a red marble is drawn from one of these five bags. What is the probability that the marble was drawn from Amanda's bag?

Let the events D_1 and D_2 stand for drawing marbles from Amanda's and Raphael's bags respectively and R for drawing a red marble.

$$P(D_1, R) = \frac{P(D_1) \times P(R, D_1)}{[P(D_1) \times P(R, D_1)] + [P(D_2) \times P(R, D_2)]}$$

Since there are a total of five bags of marbles, the probability of selecting Amanda's is 1/5 (or .20) and the probability of selecting Raphael's is 4/5 (.80). Since each bag contains ten marbles, the probability of drawing a red marble from Amanda's bag is 8/10 (.80) and the probability for the other bags is 3/10 (.30).

$$P(D_1, R) = \frac{.20 \times .80}{[.20 \times .80] + [.80 \times .30]}$$
$$= \frac{.16}{.16 + .24} = .40$$

The probability that a marble was drawn from Amanda's bag of marbles given that a red marble was drawn is .40 or 40%.

Bayes' Theorem can also be used to recalculate probabilities in light of new information. Consider the case faced by citizens in a small town down river from two large factories, one owned by Agristuff and the other by Chemofood. After several years, the citizens of the town found a high incidence of a rare cancer that had been traced to a chemical routinely used at the two companies. The town council decided to sue the company that is the most likely source of the chemical affecting their town. After an investigation, the council found that Chemofood used the chemical in 80% of its manufacturing and Agristuff used it in only 20% of its manufacturing. Based on this information, there is an 80% probability that the chemicals arriving in town are from Chemofood. Just before the council filed its action, an investigative journalist from town discovered that while Chemofood used the chemical in 80% of its manufacturing, it also had been illegally dumping 90% of its chemical waste into a stream that flows into a different watershed while Agristuff had been using the same stream to illegally dump only 10% of their waste so that 90% had been going into the river. This new information changed the council's decision to sue Chemofood.

Bayes' Theorem can be used to recalculate the probability that the chemical affecting the town came from Chemofood.

$$P(C, E) = \frac{P(C) \times P(E, C)}{[P(C) \times P(E, C)] + [P(A) \times P(E, A)]}$$

Let C be the original probability of the chemical coming from Chemofood based on its use in manufacturing (.80) and A the original probability of the chemical coming from Agristuff (.20). Let E represent the percentage of discharged chemicals sent into the town's river so that $P(E, C)$ is the probability of the chemical in the river coming from Chemofood and $P(E, A)$ is the percentage of the chemical coming from Agristuff. Given the new information:

$$P(C, E) = \frac{.80 \times .10}{[.80 \times .10] + [.20 \times .90]}$$

$$P(C, E) = \frac{.08}{.08 + .18} = .308$$

The probability that the chemical came from Chemofood turned out to be .308 and, using the principle of negation (8), the probability that the chemical came from Agristuff is $1 - .308$ or .692. The council decided to sue Agristuff.

SUMMARY PRINCIPLES

Definition of a Term and Universe

$$P(x) \geq 0 \tag{1}$$

$$P(U) = 1 \tag{2}$$

Definition of Initial Probability

$$P(x) = n \text{ such that } 0 \leq n \leq 1. \tag{3}$$

$$P(x, y) = n \text{ such that } 0 \leq n \leq 1. \tag{4}$$

Definition of Complements

$$P(x \cdot \neg x) = 0 \tag{5}$$

$$P(x \vee \neg x) = 1 \tag{6}$$

$$P(x) + P(\neg x) = 1 \tag{7}$$

Principle of Negation

$$P(\neg x) = 1 - P(x) \tag{8}$$

Exclusive Disjunction Principle

$$P(x \vee y) = P(x) + P(y) \tag{9}$$

Inclusive Disjunction Principle

$$P(x \vee y) = [P(x) + P(y)] - P(x \cdot y) \tag{10}$$

Limited Conjunction Principle

$$P(x \cdot y) = P(x) \times P(y) \tag{11}$$

General Conjunction Principle

$$P(x \cdot y) = P(x) \times P(y, x) \tag{12}$$

Definition of Conditional Probability

$$P(x, y) = P(y \cdot x)/P(y) \tag{13}$$

Principle of Total Probability

$$P(x) = P(y_1 \cdot x) + P(y_2 \cdot x) + \ldots + P(y_n \cdot x) \tag{14}$$

Limited Bayes' Theorem

$$P(x, y) = \frac{P(x) \times P(y, x)}{[P(x) \times P(y, x)] + [P(\neg x) \times P(y, \neg x)]} \tag{15}$$

Bayes' Theorem

$$P(x_1, y) = \frac{P(x_1) \times P(y, x_1)}{[P(x_1) \times P(y, x_1)] + [P(x_2) \times P(y, x_2)] \ldots + [P(x_n) \times P(y, x_n)]} \tag{16}$$

5.8   EXERCISES

5.8.1. For each of the following, indicate the interpretation operative in determining the initial probability and then compute the probabilities.

   1. If you request a *Red Hot Chili Pepper* song and the DJ randomly picks one out of the four albums she has on hand, what is the probability that it will be a song off your favorite album (assuming one of them is your favorite album and each album contains the same amount of songs).

   2. At one point, the 2003 Florida Marlins were a 1 to 160 underdog to win the World Series. At that point, what was the probability they would in fact win the World Series?

   3. The Padres have been shut out in 15 of the 100 games they have played so far this season. What is the probability that they get shut out in tonight's game?

   4. At a potluck of friends, 9 brought dishes containing meat, 4 brought vegetarian dishes, 2 brought vegan dishes and 1 brought nothing at all. If one were to randomly sample one of the dishes, what is the probability that it would be a vegetarian or vegan dish?

   5. The odds are 2 to 1 that the governor of Alaska will run for president in 2012. What is the probability of her running in 2012?

6. The 1995–96 Chicago Bulls had a regular season record of 72 wins and 10 losses. If that regular season had been extended by one game, what would have been the probability of the Bulls winning that extra game?

7. The local meteorologist on Channel 4 claimed that there was a 4 to 1 chance that it would rain this weekend. According to this information what is the probability it will rain this weekend?

8. If, out of the approximately 294 million people living in the United States, 47 million had no healthcare insurance, what would be the probability of selecting a person at random living in the United States with no healthcare?

9. The odds are 9 to 1 that Tiger Woods will win this tournament if he is ahead on the last day. What is the probability that he *loses* if he is ahead on the last day?

10. The first 43 presidents of the United States were white males. If this was all we had to go by, what was the probability that the 44th president of the United States would also be a white male?

5.8.2. Each of the following statements describes a complex event. Calculate the probabilities and show your work.

1. In a small city in a western state there is a close race for mayor and city attorney. Each race is a dead heat between a progressive and a conservative candidate. Given that no poll shows any candidate holding any sort of edge and given that the result of either race has no effect on the other, calculate the probabilities of the following events taking place, state which principles of the probability calculus you are using, and show your work.

   a. A progressive wins at least one race.
   b. A conservative wins both races.
   c. A progressive wins no seats.

2. In two adjacent congressional districts, somewhere in the United States, there is the unique phenomenon that both congressional races are a dead heat between five parties (i.e., Democratic Party, Green Party, Libertarian Party, Reform Party, and the Republican Party) going into Election Day. Given that no polls show any party holding any sort of edge, calculate the probabilities of the following events taking place, state which principles of the probability calculus you are using, and show your work.

   a. The Green Party wins both congressional seats.
   b. The Reform Party wins at least one congressional seat.
   c. The Libertarian Party wins no congressional seats.
   d. The Democratic Party and the Republican Party each win one seat.
   e. Both winners turn out to be from the same party.

3. Aurora has just finished sharing music with her friends Elena and Emma. Aurora's iPod now has 40 songs from *Rage Against the Machine*, 20 songs from *Radiohead*, 10 songs from *Tool*, and 30 songs from *System of a Down*. She puts her iPod on shuffle mode (which allows for the possibility of a song to be repeated). Calculate the probabilities of the following events taking place. State which principles of the probability calculus you are using and show your work.

a. The first song played will be a *Radiohead* song.

b. The first two songs played will be from *Rage Against the Machine*.

c. The first song played will be either from *Tool* or from *System of a Down*.

d. One of the first two songs will be from *Rage Against the Machine*.

e. The first two songs are from the same artist.

f. Redo exercise *b* but this time with a shuffle mode that does not repeat songs.

4. Aurora, Elena, and Emma bring their iPods with their shared music to a party. On their iPods, Aurora, Elena, and Emma have 40 songs from *Rage Against the Machine*, 20 songs from *Radiohead*, 10 songs from *Tool*, and 30 songs from *System of a Down*. Sarah and Christy also come to the party and bring their iPods in which they have shared music with each other. In Sarah and Christy's iPods we find 20 songs from *Rage Against the Machine*, 30 songs from *Radiohead*, 40 songs from *Tool*, and 10 songs from *System of a Down*. Assume you don't know whose iPod gets connected to the sound system. Calculate the probabilities of the following events taking place. State which principles of the probability calculus you are using and show your work.

a. What are the chances that the first song is a *Rage Against the Machine* song?

b. What are the chances that the first song is a *Tool* or a *System of a Down* song?

c. If the first song was going to be either a *Rage Against the Machine* or *Radiohead* song, what was the probability that it would be a *Radiohead* song?

d. If the first song that comes on at the party is a *Rage Against the Machine* song, what is the probability that the iPod playing is one that belongs to either Sarah or Christy? What is the probability that the iPod playing is one that belongs to Aurora, Elena, or Emma?

e. What are the chances that the first two songs are from the same artist?

f. What are the chances that the first two songs that come on will be from *System of a Down*? What are the chances if the random shuffle mode does not allow for songs to repeat?

5. In citywide elections for mayor the race is down to two candidates, one from the Democratic Party and the other from the Republican Party. In this very important election the issues of race and gender will play a huge role. For example, it is estimated that 3/4 of the Latino vote will go to the Democrat Sandra Mendoza and the remaining 1/4 will go to Republican Jack Jordan. It is also estimated that 3/5 of the male vote will go to Republican Jack Jordan and the remaining 2/5 will go to Democrat Sandra Mendoza. Enrique Perez and Leo Sandoval are both Latino males. Their chances of voting in this election are 2/3 for Enrique and 1/3 for Leo. Calculate the probabilities of the following events taking place, and in doing so state which principles of the probability calculus you are using and show your work.

a. What are the chances that either Enrique or Leo vote?

b. What are the chances that Enrique and Leo vote?

c. What are the chances that a Latino male votes for Sandra?

d. What are the chances that a Latino male votes for Jack?

e. What are the chances that Enrique votes for Sandra? And what are the chances he votes for Jack? (*Hint*: use the answers from c and d.)

f. What are the chances that Leo votes for Sandra? And what are the chances he votes for Jack? (*Hint*: use the answers from c and d.)

g. What are the chances that both Enrique and Leo vote for Sandra? And what are the chances they both vote for Jack? (*Hint*: use problems e and f.)

CHAPTER 6

PRINCIPLES OF ORDER AND DEDUCTION

6.1 INTRODUCTION

Abduction is the process, in response to an indeterminate situation, of making an assertion that has the potential to transform the situation into one that is more determinate. The assertion, "If S, then P," proposes to make sense of the initial situation by making it the consequence of some prior state of affairs. The activity of abduction itself has a similar form. Given the indeterminate situation, the act of assertion is the consequence. If the situation is indeterminate, then action is taken that tries to transform the situation into something more determinate—something with a meaning. In its broadest form "The abductive suggestion comes to us like a flash. It is an act of *insight*, although extremely fallible insight" (Peirce, 1.181). This first response (I argued in the second chapter through Peirce) is not a mere random guess, but a response conditioned by the situation and (as suggested in the discussion of Habermas) by a range of background conditions. At best, abduction can be guided by a few general commitments and a heuristic of economy (i.e., the principles of abduction proposed by Peirce). These can provide a very rough means of evaluation, but no measure of certainty.

Induction or ampliative reasoning differs from abduction because it carries out further operations that test the hypothesis while supposing determinate (or pre-designated terms), a bounded situation, and fair and appropriate sampling. These operations lead toward conclusions that are more "certain" than those of abduction. The rules of evaluation are still heuristic, but are significantly more precise than those available for abduction. Despite their increasing certainty, however, inductive arguments still have the same basic structure as the act of abduction: a series of activities framed by an if/then assertion where, if successful, actions in accord with the antecedent lead to results expected in the consequent. The connection between the instances and the conclusion is more substantial than that of a guess, but is still varied: some connections are probabilistic, some are analogical, some are based on a conception of causality, and all are the result of appropriately limited samples.

Syllogisms, in contrast, function as limit cases of certainty that include three fully determined terms. Here, valid arguments are ones where, if the premises are true then the conclusion *must* be true. Of course, such arguments still have the same overall structure as abduction. If S represents the premises of a syllogism and P the conclusion, then every valid syllogism has the form, "If S then P." As it turns out, *the ordered form of argument*—what is common among abduction, induction, and deduction—*has certain features which appear to hold regardless of particular content*. Aristotle noticed this regularity and attempted to capture the relational constraints within arguments through the theory of the syllogism. The characteristic that seemed central was validity and we used the method of Venn diagrams as a tool to prove the validity of particular argument forms. We also found out that the pattern of valid syllogisms could be understood in terms of rules that describe the features of valid forms. Using these rules or axioms, we could also show the validity of classes of arguments.

Validity in the context of induction became less clear. Lacking complete classes (either definitional or enumerative), arguments needed to be evaluated on the basis of partial classes and information. The result was a wide variety of strategies for drawing conclusions based on how information is gathered and related. While induction and probabilistic reasoning still has a stable form $(S \rightarrow P)$, we focused on evaluating how the assertions used in an argument were formed and how, in particular, they related to each other.

If we take the notion of a science of order seriously, we can go beyond the theory of the syllogism or the heuristics of induction to consider validity in its most general form as it emerges in various systems of order. Such a consideration has implications that go beyond logic to structures that govern our ordinary thinking, our politics, and even our moral judgment. As Royce suggests in his late article on order:

> The conception of order lies, therefore, just as much at the basis of an effort to define our ideals of character and society as at the basis of arithmetic, geometry or the quantitative sciences in general. ... It is, therefore, not a matter of mere accident or of mere play on words that, if a man publishes a book called simply 'A Treatise on Order,' or 'The Doctrine of Order,' we cannot tell from the title whether it is a treatise on social problems or on preserving an orderly social order against anarchy or with studying those unsymmetrical and transitive rela-

tions, whose operations and correlations upon which the theories of arithmetical, geometrical, and logical order depend. (1913, 223)

In effect, order systems pervade human experience and to the extent they are systems, there are relations and structures that are common to mathematics and moral theory, science, and social policy. Far from being a subject matter confined to an "ivory tower," Royce envisioned the science of order as making a significant contribution to human life at every level.

Royce's conception of a science of order emerged as both a general response to the metaphysics of idealism and as a particular effort to say something about the systems of geometry and logic that were emerging at the time. On one hand, the science of order would help explain in general how individuals are related to larger and larger groups and finally to the universe as a whole. The science of order thus offered a conception of diversity and unity in the most general sense. On the other hand, as mathematicians devised new systems of geometry and logicians came to see the limits of attempts to offer a single logical system as universal, the science of order provided a means to understand that no system of order could be complete and that the universe must sustain an infinite number of such systems. The science of order, in its broadest sense, came to the apparently contradictory conclusion that certain relations were necessarily universal and that the resulting system was necessarily plural. C. I. Lewis, one of Royce's students who became a leading logician, summarized the relation between the logic that came to dominate philosophy in the 20th century and Royce's logical theory, called System Σ. "Professor Royce's [logic] is the method of the path-finder. The prospect of the novel is here much greater. The system Σ may—probably does—contain new continents of order whose existence we do not suspect. And some chance transformation may put us, suddenly and unexpectedly, in possession of such previously unexplored fields" (Lewis, 1918, 369–70).

6.2   MODES OF ACTION

Most of the systems of logic devised at the beginning of the 20th century began their work by supposing the primacy of implication. Bertrand Russell, in his seminal work, *The Principles of Mathematics*, published in 1903, explained: "The relation in virtue of which it is possible for us validly to infer is what I call material implication" (1938, 33). Unfortunately, he says, "A definition of implication is quite impossible" (14). According to Russell, every effort to offer a definition still requires implication and so leaves it "fundamental." Although not a reason given by theorists like Russell, a part of the difficulty faced was that while implication was taken as fundamental, even implication required the existence of determinate logical entities for assertions to be about. Having rejected idealist concerns about the construction of things in consciousness, most logicians were left to grant an independent world of logical entities that grounded implication. Setting aside the difficulties that this ontology poses (we have already seen some of them in earlier chapters), Royce's contribution to the discussion was both an alternative orientation to logic and—crucially—an alternative approach to ontology as well. The entities with which logic is concerned are not simply given and determinate. Rather, they must be in part a product of selection that serves to mark them off from other things. In effect, prior to the process of implication there must be selection or, more broadly, a process of exclusion in which some "things" are taken to exclude other things. Only in the wake of such selective action can relations occur in the first place, including the relation of implication.

Royce's alternative approach to ontology is grounded in his proposal of a fundamental set of logical entities he called *modes of action*. We can get an idea of what a mode of action is by considering again Habermas's notion of speech acts. For Habermas, an assertion is not a determinate unit waiting to be grasped by some speaker.[1] An assertion is rather a kind of action aimed at certain purposes. By performing a speech act, the speaker uses a complex background of expectations and material resources in order to bring about a change in the hearer's dispositions, interests, or actions. As we argued earlier, an assertion amounts to a speech act in a specific situation and in its meaning amounts to a set of behaviors and expectations that will guide future action.

To say "All tigers are cats," for example, may be an action in which the speaker aims to dispose her audience to expect that encounters with tigers will involve the interactions that one finds in a diverse range of cats. Of course, it also opens a conversation about tigers and cats in which the speaker might be asked to give reasons for her assertions and in which those involved in the conversation may come to take specific action in light of claims made. Validity, for Habermas, becomes a matter of the relation of the speech act to other speech acts and the present background conditions and not a matter of some static structure independent of the situation. Whatever logical entities are involved in an assertion, they are products of a history of selection and action (not to mention evolution) and the assertion itself is also an action that stands in relation to other assertions (such as "All cats are mammals" and "All tigers are mammals") in a way that form valid arguments. Royce calls these logical entities "modes" of action in order to capture the idea that the issue in logic regards processes of ordering including those done by making assertions. "All tigers are cats" is taken up as an instance of an activity of a certain sort (a

---

[1] A standard way of thinking about utterances of language is to say that they merely indicate a "sense" or meaning (or a proposition treated as an entity). These senses are ontologically ambiguous at best. When one makes a claim about the world, she indicates a sense which is already in existence and that needs to be "picked out" from among the rest of the senses available. Frege tried to eliminate the ambiguous status of these entities by suggesting that they existed in a "third realm" accessible to the minds of speakers and hearers. One of Russell's solutions to the problem was to argue that the senses that we grasp are rather productions of our senses and are similarly caused in everyone who, for example, looks at the same object. In either case, the "entities" that an assertion means are not in any way a product of the process of selection. They are present already and are simply shared.

universal affirmative proposition) not simply as something lifted from a shelf ready-made and placed in a logic text.

Carrying the point further, it is possible to understand, as I have already suggested, that the meanings of assertions are claims about modes of action as well. To say something *is* a tiger can be taken as the claim that something acts like a tiger or equivalent to saying "it *tigers* over there." Put another way, a mode of action as known is a certain expectation about actions to come. To know a thing is a tiger is to expect certain actions. But further, to be a tiger, on this account, is to act like a tiger. To say that Barack Obama is a human being is both to have certain expectations about what he will do and to express the idea that he is, in part, the actions of a human being (even as he may also act in other characteristic ways as well). If logic is about relations among modes of action and modes of action extend beyond making claims and giving conclusions to what things are and how they relate to one another, then logic— as a science of order—can provide resources for understanding much more than arguments alone. When we return to questions of logic or order more generally, the idea that modes of action are logical entities then provides a way to understand the logical operations we have been considering.

"AGENT" ONTOLOGY

When we ask, "What is?" or "Why this?" one line of explanation is that the world is the consequence of systems without purpose. But when we try to explain the origins of purpose, it either is a mystery that enters the world from elsewhere or is not really purpose—only the appearance of purpose. An alternative approach is to begin the explanation of the world in terms of purpose. This is the approach taken by Royce at the beginning of the 20th century and by a range of philosophers at the beginning of the 21st century including Stuart Kauffman, Vine Deloria, Jr., Robert Wilson, and Karen Barad, among others.

Royce's proposal that the terms of logic be taken as modes of action also includes the idea that these modes of action are taken by *agents*, beings capable of acting with a purpose:

> It is perfectly possible to define a certain set, or 'logical universe' of modes of action such that all the members of this set are 'possible modes of action', in case there is some rational being who is capable of some one *single* possible act, and is also capable of noting, observing, recording, in some determinate way every mode of action of which [she] is actually capable, and which is a mode of action whose possibility is *required* ... by the *single* mode of action in terms of which this system of modes of action is defined. (1914, 373–4)

Royce sought to develop logical theory in a way that was consistent with the general project of idealist logic in the late 19th century. He did not, however, confine his approach to specific acts of human judgment. Royce ultimately proposed the wider notion of modes of action in order to capture purposive actions in the most general sense. While judgments can be understood as acts of affirmation and negation, Royce recognized that any action with a purpose could be characterized in the same way—as taking some course of action and negating others—even though the actions were not taken up in the form of a proposition. From this view, agency is foundational.

What is an agent? An agent is something with the capacity to act in light of what has happened and toward some possibilities and not others. By taking an action that affirms a particular belief or course of action, an agent institutes a particular system of order. This system interacts with other agents whose own purposes constrain and enable different courses of action. Such an approach does not give up the framework of a formal structure, but rather provides a framework that cannot be separated from the interests and values of the agents involved. It also ensures that any formal system that emerges from the activities of agents also can return to them in the form of the lived consequences of those activities.

Thus, if modes of action serve as the starting place for ontology, Royce proposed that certain modes of activity apply to all agential action. "*There are certain modes of activity, certain laws of the rational will, which we reinstate and verify, through the very act of attempting to presuppose that these modes of activity do not exist, or that these laws are not valid*" (1914, 365). These self-reinstating "laws" (if there are any) are necessary for any particular system of order and are such that they are created *and* discovered at once, or rather, they emerge when purposive action is taken.

One such self-reinstating principle that Royce claimed applies to all agential action is the principle of individuation: "an Individual Object is one that we *propose to regard at once as recognizable* or *identifiable throughout some process of investigation, and as unique within the range of that investigation, so that no other instance of any mere kind of object suggested by experience, can take the precise place of any one individual ...*" (1914, 350). Now, suppose that we deny the principle and so conclude that the principle of individuation is false. If it is false, then we must conclude that there are no individuals of any sort. But then our initial supposition that the principle of individuation is false violates its own claim in that it asserts something about *an individual*, in this case, an individual principle. Therefore, the principle of individuation is a self-reinstating principle.

Exclusive disjunction is also a self-reinstating principle; to see this, suppose that it is not the case that disjunction is self-reinstating. Then, of course, we assert that there are two possible states of affairs: one in which disjunction holds and another in which it does not. But these two states of affairs are distinct and related in a way that if one is true then the other is not. These states of affairs are disjunctively related, that is, by asserting that disjunction is not a self-reinstating principle, it is reinstated. Principles of logic and order emerge in the process of purposive activity. The principles of individuation and exclusive disjunction (and other such principles) are not preexisting laws for Royce; rather, they emerge in process of purposive action and provide resources for understanding inquiry and systems of order of every kind.

Traditional logics set aside agency in favor of descriptions of the products of action. There are several aspects of this approach that some take as advantages. First, the ontology is fixed, that is, the things of experience (or even of science) are what they are independent of the activities of agents. By setting aside the process by which things distinguish themselves or are distinguished by others, traditional ontologies can take things as given. Second, epistemology is determinate and "conservative." The process of knowing is collecting and ordering the products of already completed interactions in which things become individuated. Knowing conserves things as they are because the focus is on how they are "given" and not how they come to be. As Royce observed, this sort of ontology characterizes "the extreme Right wing of any elaborate social order." Such a view, he says, "has lighted the fires for the martyrs and has set up the scaffolds for the reformers" (1899, 74).

Third, values are either radically relative or given as part of the ontology of the system. Values are radically relative if they are not part of the world as given but are byproducts of the particular experience of members of the system—the human beings sophisticated enough to reflect on their interests. If the values are not simply a byproduct, then they must be given by the world itself as part of the natural order of things. In either case, responsibility for the system resides with no one (unless it is God). Individuals who challenge the system of order are either "irrational" because they undermine the system or, as "rational," perpetuate the system as given. Those who accept the idea of independent or fixed objects are "fond of insisting upon the 'sanity' of [their] views" and, in the face of criticism, "[love] to talk of 'wholesome' belief in reality, and to hurl pathological epithets at opponents ... [Such a view] is indeed the bulwark of good order" (1899, 74–5). In contrast, a conception of order that begins with agents and the structure of agency recognizes the collective responsibility of agents in the system as it develops, their role in the emergence of values, and the ongoing possibility of criticism and change.

An ontology that begins with modes of action also provides a way to understand the origins of the traditional ontologies of *realism* and *nominalism*. If one expects that the categories that result from disjunctive action are *givens* (not the result of agent action), then one adopts a realist position. From this perspective, acts of disjunction are only reports of "real" categories discovered in the world. (This leads to replacing the ontologically active exclusive disjunction with the passive inclusive disjunction.) If, on the other hand, one expects that these categories are created solely by the action of an agent, then objects are not "really" distinct at all. They are the product of naming and collecting carried out in accordance with the interests of whoever is doing the naming and collecting. This is a nominalist position. Royce rejects both extremes and can argue for his rejection in light of the nature of disjunction, which at once separates *and* connects things and agents or, given an ontology of modes of action, agents and agents.

6.3 PRINCIPLES OF ORDER

It is useful to say briefly what we mean by an object in this context. An *object* (which we will also call a "term" in this discussion) is a relation or series of relations without a separate "thing" outside these relations. Imagine, for example, the intersection of two lines. The point of intersection, strictly speaking, is the point it is thanks to the lines that intersect. To say that the point in question has an existence independent of the crossing lines is possible, of course (it would require a plane to be composed of infinitely small, indivisible points one of which is simply "picked out" by the crossing lines). Adopting Royce's approach tentatively, however, the point is something that emerges as the result of the crossing lines and with its emergence becomes an object or a term as well. On this view, objects can be understood as individuals in the ordinary sense (e.g., Socrates is an individual object or term) and in a more general sense (e.g., the class of tigers is an individual object or term). The former identifies an individual with a particular range of interactions while the latter identifies a class of individuals resulting from multiple interactions (instances of a mode of action). On this view, objects are not preexisting things discovered in the course of events, but things that emerge in the context of ordering processes including inquiry.

If we are to make sense of individuals, then we must understand what relations are as well. A *relation*, as Royce defines it, is "a character that an object possesses as a member of a collection ... and which ... would not belong to that object, were it *not* such a member." Emma is a daughter, for example, and so stands in a relation of two terms where Emma is the daughter of someone (e.g., Sarah) and we can assert "Emma is the daughter of Sarah". We can represent this relation by the expression, 'E r S', where 'r' is the relation in question and 'E' and 'S' stand for Emma and Sarah. More generally, this relation can be represented by the expression 'r(E, S)', where 'r' is the relation and 'E' and 'S' are the names of two terms that "satisfy" the relation. In general, objects (or terms) are what they are in virtue of the relations in which they stand. Emma, for example, is not only a daughter, but a student and a consumer, a bowler, and a friend. Taken together, she is at once the sum of these relations and, insofar as Emma is not confined to acting only as her categories of membership provide, she is also a unique individual who can, among other things, makes guesses that go beyond the information and experiences she has had.

If relations are the way things interact with each other, *classes* can be understood as collections of things that satisfy various relations. Classes such as human beings, students, trees, and mountains are collections of things that emerge in particular relations. At the same time, "*Relations are impossible unless there are also classes*," Royce says. "Yet if we attempt to define this latter concept, we can do so only by presupposing the conception of *Relation* as one already understood" (1914, 349). This follows since things in classes are also things interacting with each other in common ways so that being in a class is to act in characteristic ways, that is, to be in relation.

Royce proposes that classes depend upon four *conditions for their formation*. First, classes depend upon *individuation*, that is, the determination of individuals in relation. "*The concept of an individual and meaning are due to our will, to our interest, to our so-called pragmatic motives.* We actively postulate individuals and individuality. We do not merely find them. Yet this does not mean that the motives which guide our will in this postulate are wholly arbitrary, or are of *merely* relative value" (1914, 350). Individuals are not simply constituted at a whim, but emerge in the process of making sense of experience as the intersection of our interests and activity and the activities we encounter. "Without objects conceived as unique individuals, we can have *no Classes*. Without classes we can, as we have seen, define *no Relations*, without relations we can have *no Order*."

Second, classes depend on the principle of exclusion or *negation*. In order to define any individual or group, the act of definition includes both an affirmation of the individuals and the negation of other things that are different or outside the individual or group affirmed. If x is a class, then the principle of exclusion posits another class, ¬x, that is, the *complement* of x and includes all those things (or activities) that are not x.

By establishing two classes a third principle emerges as well called *inclusion* or *subsumption*. Now suppose that two classes, x and y, are created. These classes may or may not completely exclude each other, but are nevertheless well defined. In addition to these two classes, two more are necessarily generated as well. The first, called the logical product of x and y, includes everything that is a member of both x and y. The second class, called the logical sum, includes everything that is a member of either x or y. Since individuals can be members of classes, subsumption sets up the possibility that classes will be connected and divided in certain ways. This principle is easily seen in the theory of the syllogism as underlying the relations between the terms. If all tigers are cats and all cats are mammals, the syllogism presents a situation in which tigers are included in cats and cats in the class of mammals and so tigers are included in the class of mammals.

Finally, each individual determined in the context of a situation has the potential to be associated with one or more classes. This relation, Royce called it *belonging*, is the lynchpin for understanding order and an affirmation that classes and relations are necessarily bound together in the processes of ordering. Since a given individual can belong to a class, its relations can be evaluated and its activities ordered. If a thing appears that does not belong to a recognized class, then we do not know what to expect. One response to such indeterminacy is to inquire into the thing and determine the class to which it belongs. If we can find nothing to which it belongs, we propose a new class and begin to set new expectations for our interactions with it.

Behind these four principles is a fifth, *the principle of continuity*, that is, the idea that everything is connected in a continuous field out of which the particularities of experience develop. Such a view is the antithesis of the idea that the universe is fundamentally "atomic," that is, made up of indivisible, independent particles of the sort posited by many 20th century logicians. In this discussion, we will set aside the most basic questions of metaphysics, not because they are irrelevant to the science of order and so to logic, but because they are so important that they merit extensive treatment impossible in the context of an introduction to logic. Regardless of these limits on discussion, however, we can nevertheless identify important implications of continuity for our present discussion. One result is that the role of the inquiring agent is crucial to "what there is"—that is, to the existence of the world of things and events. The continuity of our starting place, however, also ensures that the inquiring agent cannot act independently since what is individuated by the agent is also acting and so individuating itself.

6.4   LOGIC AND THE ACT OF JUDGMENT

In the 19th century, idealist logicians understood logic as a study of acts of judgment. These acts require the participation of conscious agents who, when faced with a choice, are required to decide. Unlike the logic that later emerged as a formal means of understanding given terms in relation, idealists were interested in first understanding the role of agents in the process of generating the individuals and categories of experience. Since judgment was at the center of logic, it also was taken as a study that could shed light on moral as well as other forms of judgment. Rather than stripping logic of its relevance to experience, idealist logicians sought to make logic foundational in the study of lived experience.

At its simplest, a choice or judgment is a relation between affirming or not affirming a particular belief or plan of action. When faced with the choice between believing a friend and not believing or meeting a friend for coffee or doing something else, the relation one faces can be represented (for example) as 'p | ¬p' (read "p slash not p"), where 'p' represents a particular instance of believing or meeting or knowing, '¬p' as *not* taking the action, and '|' as the relation between the two alternatives.[2] The expression indicates a situation of choice in which one alternative or the other will be taken (will be true) but both cannot. Both alternatives, p and ¬p, are available to an agent and the alternatives are incompatible or exclusive in the sense that taking one excludes the other. Call this basic two-term relation *exclusive disjunction*.

Any action serves the function of excluding a range of actions when it is taken. To make a left turn, is to rule out taking a right at that moment. To sit rules out running at the same time. To make an assertion rules out saying nothing. Hollering an insult blocks giving a compliment. Asking a question overrides disinterest. There are, of course, other actions that are not excluded when one turns or sits or asserts or insults or questions. These other actions— chewing gum while walking, speaking truly while asserting, and so on—are actions that are compatible with the one taken. In fact, one way to understand an act is to see it as defining a *universe of discourse* where there are a range of actions that can be taken and a range that cannot be taken.[3]

---

[2] The vertical stroke, '|', is also used in some logic texts to symbolize what is called the "Sheffer Stroke." The relation I represent here with the stroke differs from the Sheffer relation in that the Sheffer relation holds if one or the other term is true or neither is true, but not both. Henry Sheffer, who originated the use of this relation, was a renowned logician at Harvard and had been a student of Royce. It seems likely that his formulation of logic in terms of this disjunctive relation is, in part, a product of his work with Royce. Russell and Whitehead recognized the importance of Sheffer's work in the revised edition of *Principia Mathematica* in which they admit that their system could be expressed in terms of the stroke relation alone.

[3] One can understand Royce's notion of a universe of discourse rather broadly following an early meaning of "discourse," "to run, move, or travel over a space, region" (OED). Such a universe, defined by some initial action literally defines a region of action or "of travel" in terms of which expectations and actions are organized. Also see Section 5.7.

This relation is a peculiar (though familiar) one. What it represents in experience is a moment of suspense in which one recognizes (or determines) that she has an exclusive choice to make. It also marks a moment in which to act at all is to take one of the two alternatives. While there are certainly times when one can choose both alternatives ("Do you want cake or ice cream?" "I'll have cake *and* ice cream!") or neither ("No thanks, I won't have dessert"). The relation described here is one in which the choice is required and one does not have the options of choosing both or neither. Such a moment is characterized by incompatible courses of action, but also as a moment in which, like it or not, something will happen. In the context of his discussion of the "will to believe," William James argued that this characteristic is part of what makes alternatives "forced options." In such cases, one can choose from the available alternatives, but not choosing counts as an action as well: "if I say, 'Either accept this truth or go without it,' I put on you a forced option, for there is no standing place outside the alternatives" (1897, 3). In response to James Clifford's injunction to proceed skeptically when faced with such options, James concludes "if we believe that no bell in us tolls to let us know for certain when truth is in our grasp, then it seems a piece of idle fantasticality to preach so solemnly our duty of waiting for the bell. Indeed we *may* wait if we will ... but if we do so, we do so at our peril as much as if we believed. In either case we *act*, taking our life into our hands" (1897, 30).

The moment of choice represented by 'p | ¬p' is a *modal relation* or one that has to do with necessity and possibility. In the case of a choice between doing A and not doing A, it is clear that doing A necessarily rules out the possibility of not doing A. While this relation seems apparent of exclusive disjunctions like A | ¬A, the same basic structure holds for other incompatible actions as well. Suppose that Sophie is faced with two alternatives: to study or to take the evening off. Formally, the relation of choice is between studying and relaxing (say S and R) and, since these are incompatible S | R. In this case, we might not want to say that she must choose between the two alternatives (since she could, for example, work out at the gym which, at least for some, is neither studying nor relaxing). While p | ¬p is an exclusive disjunction in which one or the other alternative must be taken, if any action is, S | R seems to represent a pair of options that may both be rejected.

Yet the presence of other alternatives, rather than marking something that eliminates the "forced" character of the situation, instead marks a more complex choice that remains a forced option. If Sophie adds some alternative P to the disjunction S | R, taking P may rule out S (if it is compatible with R) or rule out R (if it is compatible with S) or it may rule out both (if it is compatible with neither S nor R). At the same time, not making a choice is (following James) an option that also amounts an action so that if P is not making a selection, then it is an option exclusively related to S and R. Notice that relative to P, S and R, taken together are equivalent to ¬P in that if either is taken, then P is not. Regardless of how many additional options are added to the original moment of choice, the relation remains exclusive and marks a necessary relation between the alternatives.

The central relation of judgment then is a relation between exclusive alternatives or *disjuncts* such that, from the perspective of the original situation of choice, one of the disjuncts (or an action equivalent to one of the terms) must be taken. If this analysis is correct, then a third element of the relation should be apparent as well. The disjunction is not simply the relation between two independent terms set out before a person in experience or that just happen to be disjunctively related, rather they are the product of a rich context in which the disjunction has emerged. The third term, captured by the symbol '|', can be understood as something that is *between* the alternatives that is compatible with either one but is not compatible with both. If all the terms in question are modes of action, as I proposed earlier, then what is between the terms of an exclusive disjunction is the act of division or what is sometimes called a *cut*. Sophie's initial choice of studying or relaxing stands in a moment of experience that is conditioned by her interests and circumstances, her relation to others, and so on. The cut does not represent a passive set of circumstances, but *the presence of an agent for whom the choice is a choice and whose action (or failure to act) will resolve the suspended moment at hand.*

The cut that sets up the moment of choice is illustrated as well by the traveler we met in chapter 2. Suppose the traveler, Jane, is on a walk in the forest and, at a certain point, she reaches a fork in the path. While both the left and right forks are open to her, she cannot take them both. After a pause, she takes the left fork and this requires that she not take the right. The situation of judgment seems automatic: choose left or right. But this masks the process. The exclusive disjunction that marks judgment is a product of a situation *and* the action of an agent, Jane, who recognizes (or takes) the situation as a choice between the possibilities of going left or right. One might argue that Jane is only incidentally involved and the mere fact of the fork in the road establishes all that is necessary to require that choosing the left fork requires not choosing the right. In terms of inquiry, the fork in the road is an indeterminate situation and makes a range of possibilities available. It takes Jane's action to settle the issue by taking the problem as one of choosing left or right. But this does not settle everything—it opens a new range of possibilities that will be encountered by the agent as she proceeds.

The suspended moment of S | R marks the possibility of four alternative outcomes (in this case, two different choices and their contrapositives). Suppose, for example, that Sophie rejects the idea of relaxing and so it follows that she will study or do something compatible with that choice. She might, for example, choose to improve her chances of getting a good grade on an exam and so will take actions compatible with her choice. Studying is one of the actions compatible with the goal and so is, in this sense, equivalent to S and incompatible with R. The original choice, then, sets up a conditional situation. If Sophie rejects taking the evening off and all things compatible with it, then she sets out on a course of action to do things that are compatible with the choice of studying. But can't Sophie reject studying as well in favor of some other action, say going to bed early or washing dishes? She can, of course, but this marks another moment of choice, this time between continuing on the course of action compatible with study or rejecting it in favor of another that is compatible with doing dishes, S | D, or some other alternative. This does not change the character of the moment of choice, but suggests that the relation does not describe a permanent state of affairs, but rather a moment of experience.

The four possibilities that follow from S | R can be summarized this way: given S | R, it follows that if not-S, then R and if not-R, then S. Since S | R describes a forced option, it is also equivalent to ¬R | ¬S, and so it is also the case that if ¬¬R (or not taking not-R, that is, taking R), then not-S and if ¬¬S, then not-R.

There is, however, a problem. This set of inferences from the original disjunctive situation was at the center of one of the key challenges to idealist logic at the beginning of the 20th century and in part responsible for setting aside this approach to logic. In the situation described above, given S | R, it follows that if ¬S then R. But again, given that S | R, it also follows that if R is taken then ¬S follows. The result is an argument of the form

$$\neg S \rightarrow R$$
$$R$$
$$\therefore \neg S$$

This, the critic G. R. T. Ross observed, is simply the *fallacy of affirming the consequent* we saw in Chapter 3. The same relation, S | R, also leads to an argument of the form if S then ¬R and ¬S, therefore ¬¬R. In this case, the exclusive disjunction appears to lead to the *fallacy of denying the antecedent*. These results led Ross (and many others) to conclude that the exclusive disjunction is not, in fact, a basic function of logic but a relation better understood in terms of other operators.[4]

Ross was correct in understanding that the exclusive disjunction leads to the fallacies described, but only if the original relation is understood as a description of a static relation between terms. The relation described holds that either Sophie will take S (she will study) and so ¬R (will not relax) or she will take ¬S (not study) and so R (she will relax). The fallacies occur only if we suppose that she can do both. While both follow from the original situation, the point of the relation that describes choice is that what follows will be the result of a judgment. Rather than setting out a universe of fixed entities, the exclusive disjunction marks *diverging paths* where what happens next will be a product of an agent making a choice. Exclusive disjunction describes a forced option in James's sense in which an agent, Sophie, for example, takes an action and in so doing leaves behind the situation S | R and replaces it with a new unified situation in which either S and ¬R or ¬S and R occur.

When faced with a moment of choice, Sophie has yet to act, but the situation at hand is characterized by four patterns of action each of which have the potential to rearrange the present situation and lead to a new unified one.

---

[4] These operators are inclusive disjunction, negation, and conjunction.

In this case, the conditional proposition expressed "if S then not-R" is a hypothesis in which, if S is taken or believed or true, then not-R will follow. But why isn't this just the same course of action as the one described if not-S then R? Since S and ¬R in fact are equivalent, it would appear that the first conditional proposition simply says that if S then S and the second says (since ¬S is equivalent to R) that if R then R. While the equivalence is true relative to the initial situation of choice, the two courses of action are different in where they lead. When Jane faces the choice of two paths, the selection of one or the other is sufficient to the situation. However, the two paths lead to different places and so, while the choice amounts to affirming one path or the other, what one sacrifices and inherits through the act of choice will be different (if only because S, even if it is identical with itself, is also incompatible with R). In anticipation of these divergent courses, the conditionals proposed amount to hypotheses about what will follow from the alternatives at hand.

6.5 DEDUCTION: THE LOGIC OF ASSERTIONS

Until this point, we have examined validity by a variety of means including examination of background conditions, Venn diagrams, and axiomatic truths. By introducing the notion of ordered relations among modes of action, we can add a fourth approach to validity. Following the structure just suggested, a proof of validity can be one in which we begin with the description of a certain situation (that, for example, p and q are incompatible alternatives). An argument is valid if we can arrive at a conclusion by means of a series of steps where each step follows from the situation given. We can examine validity from this angle by establishing a system that represents the basic relations of order that we have already discussed. Using the observations just made, a set of symbols representing possible actions and relations, and a set of rules relating these symbols, we can establish a system that, like the theory of the syllogism, will allow us to examine the validity of arguments. The symbols and rules that we develop in this chapter regard relations among propositions and so the *terms* of this formal logic are complete assertions and not parts of assertions (that is, subjects and predicates). The relations among modes of action (e.g., choosing among alternatives) hold as well for assertions, which are simply one form of action. In general, the system we give here can be called the logic of assertions or, more commonly, propositional logic (in contrast to the predicate or quantificational logic in the next chapter).

To begin, *let lowercase letters p, q, r, s, ...* represent distinct modes of action. For our purposes, these distinct "modes" will represent the propositions expressed by particular assertions.

If p and q are propositions, then the following are also propositions:

¬p   "not p"

◇p   "it is possible that p"

□p   "it is necessary that p"

p	q	"p and q are exclusively related"
p ∧ q		"p and q"
p ∨ q		"p or q" (*Inclusive* disjunction)
p ⊃ q		"if p then q"
p ≡ q		"p if and only if q" (Biconditional)

Parentheses are used as a convenience to mark individual terms and should be used to indicate the relation between two terms. For example, if 'p ∧ q' is joined with r by a '∨' then '(p ∧ q) ∨ r'.

Let uppercase letters A, B, C, D, ... represent particular actions or, for our purposes, particular assertions. "All tigers are cats" and "Grass is green" are assertions indifferently represented by 'p'. However, "All tigers are cats" can be represented as a particular assertion as 'A' or any other uppercase letter.

Let p | ¬p represent the fundamental relation, a situation characterized by a "forced option," where '|' is the symbol for exclusion and '¬' is the symbol for negation. When '¬' is added to the symbol for an action (proposition, or particular assertion) it indicates that the action is not taken (or it asserts the complement of the original proposition).

Let two terms be equivalent if and only if they can be substituted for each other in any relation in an argument without changing the relations expressed by the proposition. If p and q are equivalent, this can be represented 'p = q'. If p and ¬p are incompatible and q = ¬p then p | q where q replaces ¬p in p | ¬p.

Let p ∧ q represent the conjunction of two actions or propositions. To assert that p and q is to assert that both actions are taken.

The exclusive disjunction is a modal relation. Since p | ¬p is a situation in which it is not possible that p and ¬p can both be taken *and* one of the disjuncts must be taken. The first half of this expression can be represented '¬◇(p ∧ ¬p)' or "it is not possible that p and not-p can both be taken," where '◇' is the symbol for "it is possible." If q is not compatible with p, then ¬p = q and so p | q can also be represented '¬◇(p ∧ q)', "it is not possible that p and q can be taken together." The symbol '□' can be used to express "it is necessary that" and so '¬◇(p ∧ ¬p)' is equivalent to □¬(p ∧ ¬p), "it is necessarily not the case that p and not-p can be taken together." The second half of the meaning of the exclusive disjunction is □(p ∨ ¬p) or "it is necessary that p or not-p is taken." If q = ¬p, then □(p ∨ q).

The following are *rules of equivalence*:

Exclusive Disjunction:	p	q = ¬◇(p ∧ q) ∧ □(p ∨ q)
Necessity (N):	¬◇p = □¬p

Possibility (P): $\neg\Box p = \Diamond\neg p$

Double Negation (DN): $p = \neg\neg p$

Commutation (C): $p \mid q = q \mid p$

$p \wedge q = q \wedge p$

$p \vee q = q \vee p$

$p \equiv q = q \equiv p$

Association (A): $p \wedge (q \wedge r) = (p \wedge q) \wedge r$

$p \vee (q \vee r) = (p \vee q) \vee r$

Exclusive Disjunction expresses the meaning of the relation discussed in the previous section as the original situation of judgment. Rule N relates necessity and negated possibility while Rule P relates possibility and negated necessity. The rule DN expresses the idea that not failing to take an action (that is, to not not-take an action) is simply to take the action. If this is so, we can replace the term "p" with the term "$\neg\neg p$" and vice versa without changing the relation expressed. The rule C represents the idea that exclusion, conjunction, inclusive disjunction, and biconditionals are symmetrical relations. If two actions are both taken, then both occur together and the order in which they are represented does not matter. The third rule, A, simply expands C and recognizes that order is unimportant for conjunction and inclusive disjunction.

The basic relations of the system lead to certain other relations. When given the possibility of two incompatible actions, for example, and one of them is actually taken, then a new situation is entailed in which the option not taken is negated. Four such entailments are proposed for the relations of $\mid$ and $\wedge$, exclusion and conjunction.

These are *postulates of the system*:

Exclusion (Ex): $(p \mid q) \wedge p \Rightarrow \neg q$

$(p \mid q) \wedge q \Rightarrow \neg p$

Simplification (Si): $p \wedge q \Rightarrow p$

$p \wedge q \Rightarrow q$

Elimination (El): $(p \mid q) \wedge (\neg q \mid r) \Rightarrow p \mid r$

Material Implication (MI): $p \mid \neg q \Rightarrow p \supset q$

$p \mid \neg q \Rightarrow \neg p \supset \neg q$

Here, '$\Rightarrow$' represents the asymmetrical relation of entailment. The postulate of exclusion (Ex), says that when p excludes q and p is taken, then q is not taken. The relation is a necessary one (as discussed above). Simplification (Si) states that if p and q are both actions taken, then the situation entails that p is taken. Elimination (El) makes it clear that the exclusion of one relation can interact with the exclusions of another. The explanation is straightforward. Suppose that we have two actions that exclude one another, for example, Juan can go to a movie or stay home and read. Now suppose that by not staying home, Juan will miss seeing his cousin when she drops by. The postulate of elimination states that given the first two actions (going to the movies and staying home) and the second two (not staying home and seeing someone), it is entailed that Juan's going to the movies also rules out seeing his cousin. The middle term, "staying home" can be "eliminated" from the actions specified because the first term is found to be incompatible with the third.

The fourth postulate, *material implication*, is entailed by the exclusive disjunction, but is not equivalent to it. If the actions p and q are related as the left side of the entailment suggests, that is, p excludes not taking q, then it follows that if the action p is taken, then action q must also be taken. Note that *two* material implications are entailed by the initial situation (i.e., the result that led Ross to reject exclusive disjunction). However, the two implications cannot both be taken (one supposes p and the other supposes ¬p). What we can do is to try the alternatives that follow from the initial situation and determine what might happen in light of the judgment made. The material implications, in this sense, are hypotheses that enable inquiry to go forward. The initial situation sets the problem (to choose between options) and hypothetical reasoning allows one to test the possibilities before taking overt action. *In the context of proving validity only one of two possible material implications can be used.* To use both would suppose that two incompatible actions can both be taken.

Given the initial situation, I might hypothesize: if I go to the basketball game, I will see my team win. This is surely the case if going to the game somehow excludes a loss. If it doesn't (and this seems likely), the hypothesis that follows is one that will hold in three different circumstances. (1) If I go to the game and my team wins, then the hypothesis would be confirmed. (2) If my team wins, then, again, it must be true that if I go, they will win. And, (3), the hypothesis continues to hold as long as I do not go to the game and the team does not win. The only case in which the hypothesis does not hold is if I go to the game and my team loses.

Suppose that we layout the possibilities expressed by the conditional, "if I go to the game, my team will win" represented as 'p ⊃ q'. We can present this as a table (Fig. 6.1).

p	q	p ⊃ q
T	T	T
T	F	F
F	T	T
F	F	T

Figure 6.1   Truth Table for Material Implication

Let 'T' stand for "the action is taken" and 'F' for "the action fails" or "the action is not taken."[5] The conditional is a viable expression of my options (of actions that can be taken) if I both go to the game and my team wins. It also holds if, though I do not go, my team wins (after all, I could have gone). Finally (and a bit strangely), the conditional holds if neither thing comes to pass, since neither of the actions involved were actually taken, it leaves the expression of options unfulfilled. If, however, I go to the game and my team loses, the conditional turns out to not hold at all.

The initial situation marks what I take to be a state of affairs that connects my attendance and the success of my team. The material implications that are entailed by my initial situation propose for consideration how the evening might play out. If I try the first range of possibilities (if I go to the game, then my team will win), it will hold in every case except the one in which I attend the game and the team loses. The other range of possibilities ($\neg p \supset \neg q$) will hold in all cases except the one in which I do not attend the game and my team wins. Since the two alternatives are incompatible, I will eventually go or not and my hypotheses will be confirmed or overturned.

Logicians have discussed material implication at length. It has a number of advantages as the fundamental logical operation even as it requires that we make a number of suppositions regarding both the nature of logic and the nature of the world. The logic of material implication begins by recognizing the situations described above in which material implication, $\supset$, seems viable. These can be summarized in terms of a new logical operation called *inclusive disjunction*. An inclusive disjunction, symbolized 'p ∨ q', and read "p or q" appears to be similar to the exclusive disjunction that began our discussion. It is not the same, however, since the expression "p ∨ q" holds when either p is taken or q is taken or *both* p and q are taken. If this is the meaning of the inclusive disjunction, then we can actually redefine material implication in terms of ∨. We can represent a table for inclusive disjunction as shown in Figure 6.2.

p	q	p ∨ q
T	T	T
T	F	T
F	T	T
F	F	F

Figure 6.2   Truth Table for Inclusive Disjunction

[5] Traditionally, in tables like this, 'T' and 'F' represent true and false. We will stay with the language of action here in order to emphasize the centrality of action in the processes of order. None of the results, however, will differ if one chooses to talk in terms of truth and falsity as long as it remains clear that true and false regard whether or not the actions in question are taken.

p	¬p	q	¬p ∨ q
T	F	T	T
T	F	F	F
F	T	T	T
F	T	F	T

Figure 6.3   Truth Table for Material Implication Using Inclusive Disjunction

The inclusive disjunction fails to hold only when neither of the disjuncts is taken. If you consider the table for material implication again, it is clear that the expression, p ⊃ q, holds whenever q is taken *or* when ¬p is taken (i.e., p is negated). From this perspective, it follows that p ⊃ q = ¬p ∨ q. If we adjust the table to represent the relation between ¬p and q then we get the truth table shown in Figure 6.3.

Notice that the column representing p, q, and the disjunction are the same as the values of the column representing the values in the table for material implication (Figure 6.1). These two expressions, p ⊃ q and ¬p ∨ q, hold under exactly the same circumstances and so they are equivalent.

We may add a final relation as well. Sometimes we assert that two terms are related in such a way that they imply each other. For example, Joe might inherit his family's millions if he graduates from college (or, slightly restated, "if Joe graduates from college, then he will inherit his family's millions"). It also might be true that the trust fund documents are written in such a way that if he does inherit his family's millions, then he will have graduated from college. From this perspective, we can say that Joe inherits millions if and only if he graduates from college. This relation is called a biconditional and is represented by '≡'.

STRICT IMPLICATION

The original situation of choice actually implies a relation called *strict implication*. This relation, symbolized 'p < q' ("p strictly implies q"), expresses a necessary relation between p and q, that is, □(p ⊃ q). A student of Royce's, C. I. Lewis, first proposed the strict implication relation as a way to address what he called the paradoxes of implication.

In the logical system devised by Russell and Whitehead, the central operation is material implication, a relation that is true if both terms are true, if the antecedent is false and the consequent is true, or if both terms are false. As a result, a material implication is always true if the antecedent is always false or the conclusion is always true. Put another way, any false proposition implies a true proposition and any true proposition is implied by any proposition, true or false. So, from the perspective of material implication, we can say that if there are square circles, then Barack Obama is president of the United States. And we can also say that if pigs can fly, then most Americans oppose the war. These expressions are paradoxical because material implication allows the antecedents and consequents to have nothing to do with each other. Even though both propositions are formally true according to the truth table of material implication, the antecedents and consequents are unrelated so that, contrary to our expectations about if/then propositions, the truth of one has no bearing at all on the truth of the other or of the whole. Strict implication solves this problem by requiring a necessary relation between the terms.

Lewis's work on strict implication led to the development of modal logics that can take the requirement of necessity into account. It also inspired work on more recent non-classical logics including so-called "relevant" logics that require an even greater degree of connection between the terms. Relevant logics also have the possibility of addressing the new paradoxes that follow from strict implication. These include the fact that any proposition that is necessarily false implies every true proposition and any proposition whatever implies any necessarily true proposition.[6]

The exclusive disjunction $(p \mid q)$ as it is developed here is equivalent to the conjunction of two modal statements: $\neg\Diamond(p \wedge q) \wedge \Box(p \vee q)$. Propositions of this form entail two different strict implications: $p < \neg q$ and $q < \neg p$. If these were to hold simultaneously, then, as discussed earlier, the exclusive disjunction would lead to the fallacies of affirming the consequent and denying the antecedent. However, if the original situation is one of forced choice, then only one of the strict implications can be taken. If this is the case, then $p \mid q$ entails $p < \neg q$ (for example) and the strict implication (using a standard modal logic) entails the material implication $p \supset \neg q$.[7] Since the original situation entails a system of order based on strict implication and this system implies the system of material implication, the system of order that uses material implication is a proper subset of the system of order that uses strict implication, which, in turn, is a proper subset of the system founded on exclusive disjunction. It is also true that since the original situation entails two incompatible strict implications, an act of judgment is required to select one course of action or the other. Similarly, the system, of material implication requires an act of judgment to select the hypotheses to be considered in light of the original situation. The resulting formal system proposed here recognizes the role of agents, the process of inquiry, and the emergence of standard logics as a subset of a system of order that begins with acts of judgment.

We can now add some further equivalent relations:

Disjunctive Implication (DI)	$p \supset q = \neg p \vee q$
Contraposition (CP)	$p \supset q = \neg q \supset \neg p$
Biconditional (B)	$p \equiv q = (p \supset q) \wedge (q \supset p)$

[6] See Graham Priest, *An Introduction to Non-Classical Logic* (2008) for a discussion of Lewis and later developments.

[7] Lewis gives a proof that material implication is entailed by strict implication (Lewis and Langford, 1959, 140). Using the modal language S5 (or Κρστ) and the proof procedure of tableaux in Priest (2008), it can be shown that $p \mid q$ entails the strict implications, $p < \neg q$ and $q < \neg p$. In proving material implication for this system, one must suppose that only one of the strict implications is taken.

De Morgan's Laws (DM) $\neg p \vee \neg q = \neg(p \wedge q)$
                                $\neg p \wedge \neg q = \neg(p \vee q)$

Distribution (D) $p \vee (q \wedge r) = (p \vee q) \wedge (p \vee r)$
                                $p \wedge (q \vee r) = (p \wedge q) \vee (p \wedge r)$

De Morgan's Laws—a product of the work of Augustus De Morgan in the mid-19th century—illustrates the relation between conjunctions and inclusive disjunctions. Consider $\neg p \vee \neg q$. This assertion means that either p is not taken or q is not taken or both are not taken. The only situation it excludes is one in which both p and q are taken. Put another way, $\neg p \vee \neg q$ means that there are *no* cases in which *p and q are taken together*, that is, it is not the case that p and q. Therefore, $\neg p \vee \neg q$ means the same thing as $\neg(p \wedge q)$ and so the terms are equivalent. This can be illustrated by beginning with the conjunction as well. If it is not the case that p and q are taken together, then it is the case that p is not taken or q is not taken or both of them are not taken, that is, either not-p or not-q or both (but not neither). A similar informal proof can be given for the second De Morgan Law. Beginning with the disjunction, to say that it is not the case that p is taken or q is taken or both are taken, it must be the case that p and q fail together, that is, it is not the case that p and it is not the case that q. The other laws can be explained in a similar fashion or illustrated by means of a table of values of the sort we used above.

Combining all of the equivalences and postulates above, we are in a position to evaluate the validity of arguments in light of the relations between the terms of an argument. These proofs of validity require that the assertions considered be presented in a standard form, in this case, one where all of the symbols are properly used and there are no undefined variables. A *proof of validity* will be an argument that consists of a statement of the premises given in standard form followed by a series of further propositions (also represented symbolically), each justified by a postulate, an equivalence, or another valid argument form.

To use a postulate in a proof, you must find propositions of the form shown on the left side (or antecedent) of the postulate. You may then assert the right side (consequent) of the postulate on a later line of the proof and justify it by listing the line numbers that represent the antecedent and the abbreviated name of the postulate. For example, if we are given the relation between two assertions, A | B, and an assertion that one of the possibilities is taken, A, we can validly conclude that $\neg B$. This is easily shown in the form of a proof.

1. $(A \mid B) \wedge A$ Premise
2. $\neg B$ 1 Ex

To use an equivalence in a proof, you may substitute equivalent expressions wherever they occur in a line by asserting a new line with the equivalent in place. The line is justified by giving the line number of the original assertion and the abbreviation for the equivalence used. So in the following, we want to show that given A | B, we can conclude a material implication, A ⊃ ¬B.

1. A | B   Premise
2. A | ¬¬B   1, DN
3. A ⊃ ¬B   2, MI

Given the form of material implication, p | ¬q ⇒ p ⊃ q, the procedure is to preserve the order of the exclusive disjunction (in this case, 'p ... q'), replace '|' with '⊃', and drop '¬' from the term on the right side. Since 'B' in the premise does not have a '¬' to drop, we add two by replacing 'B' with the equivalent term, '¬¬B' and justify the replacement by listing the original line (1) and the justifying rule (DN). The conclusion then follows using line 2 and the postulate MI.

To use a valid argument form to justify a line in a proof requires that we have some valid forms. The simplest argument form is the *Disjunctive Argument.* We are given the assertion, A ∨ B, where ∨ means that either A is taken or B is taken or both are taken. We are also given ¬A, that is, A is not taken. It follows that B must be taken, since, if p ∨ q holds and p is not taken, q is required. This gives us a general argument form of the following:

$$p \lor q$$
$$\neg p$$
$$\therefore q$$

This valid argument form can be used in a proof if assertions that fit the form of the premises of the argument appear in the proof. We can assert the conclusion of the argument on a later line.

A second useful valid argument form is *modus ponens*: given the conditional, if p then q, and the assertion, p, we can conclude that q. This form can itself be proved using the resources now available:

1. p ⊃ q   Premise
2. p   Premise
3. ¬p ∨ q   1, DI
4. ¬¬p   2, DN
5. q   3, 4 Disjunctive Argument

The argument form, *modus tollens*, is closely related. In this case, if we are given the conditional, if p then q, and q does not hold (that is, not-q does hold), then we can conclude that not-p. The proof for this argument form follows the same pattern. Notice that lines 2 and 4 are out of the usual order for disjunctive argument. Line order does not matter since the premises are connected by the symmetrical relation "and."

1. $p \supset q$ Premise

2. $\neg q$ Premise

3. $\neg p \vee q$ 1, DI

4. $q \vee \neg p$ 3, C

5. $\neg p$ 2, 4 Disjunctive Argument

Conjunction is an argument form that simply combines two lines into one by using the conjunction sign. If we have p on one line of a proof and q on another, then both are asserted (both are taken) and so we may also say that $p \wedge q$.

A *conditional argument* is one that constructs an implication using an assumption and the resources given in the premises (if any). Suppose that we would like to show that, given the premises $p \supset q$ and $q \supset r$, we can conclude that $p \supset r$. To show this we can assume p and if we can arrive at r using p *and* one or more of the premises, then we can assert $p \supset r$ by Conditional Argument.

1. $p \supset q$ Premise

2. $q \supset r$ Premise

    3. p   Assume

    4. q   1, 3 Modus Ponens

    5. r   2, 4 Modus Ponens

6. $p \supset r$ 3-5, Conditional Argument

The conclusion lists the range of lines used in the conditional proof and the lines indicate this range as well. The inset line numbers indicate the assumption and the premises of the argument. The resulting proof, in this case, is of the argument form called *transitivity*, that is, given if p then q and if q then r, we can conclude that if p then r.

Another useful form of argument is called *reductio ad absurdum*. These arguments are based on the idea that if a series of deductions leads to a contradiction (an assertion of the form $p \wedge \neg p$, one of the premises is mistaken and should be replaced by its negation. To use this form, we may introduce *an assumption which is the negation of the conclusion we would like to prove*. This argument form captures the idea that arguments are a means for testing in thought possible courses of action. The form asks, in effect, is a particular outcome compatible with the premises that I take to define a possible course of action? If the combination of the premises and the complement of the expected conclusion lead to a contradiction, then this proves that the complement of the conclusion is incompatible with the premises and so the original conclusion (the complement of the complement) must be compatible. For example, one of the consequences of De Morgan's Law should be that if we have ¬p ∨ ¬q as a premise, we should be able to derive ¬(p ∧ q) (and not simply by using the equivalence rule). Given ¬p ∨ ¬q, assume the negation of the desired conclusion, that is, ¬¬(p ∧ q). If this assumption, in combination with the premise, leads to a contradiction, then we know that the negated conclusion fails (or is incompatible with the course of action defined by the premises) and we know that the negation of the assumption must hold.

1. ¬p ∨ ¬q	Premise
2. ¬¬(p ∧ q)	Assume
3. p ∧ q	2, DN
4. p	3, Si
5. ¬¬p	4, DN
6. ¬q	1, 5 Disjunctive Argument
7. q	3, Si
8. ¬q ∧ q	6, 7 Conjunction
9. ¬(p ∧ q)	2-8, Reductio ad absurdum

In this case, the inset line numbers indicate the assumption and the assertions that are entailed by the assumption and the premises including the contradiction (line 8 in this case) that indicates that the assumption is mistaken. When a contradiction is reached and stated, on the next line we can list the assumption without its negation and justify it by giving the range of lines from the assumption to the contradiction and listing *reductio ad absurdum* as the valid argument form we used. The proof shows that ¬(p ∧ q) follows from ¬p ∨ ¬q, and we can also prove that ¬p ∨ ¬q follows from ¬(p ∧ q).

Addition is an argument form based on the inclusive disjunction. Given any assertion, p, it follows that p ∨ q must hold since at least one of the two terms is taken (p according to the premise) and it does not matter if the other term holds or not.[8]

In sum, the valid argument forms that can be used in proving validity are as follows:

Disjunctive Argument (DA):   p ∨ q
                              ¬p
                              ∴ q

---

[8] This argument form makes apparent the paradoxes involved in material implication discussed in the section on strict implication above.

Modus Ponens (MP):

$p \supset q$
p
$\therefore q$

Modus Tollens (MT):

$p \supset q$
$\neg q$
$\therefore \neg p$

Transitivity (TR):

$p \supset q$
$q \supset r$
$\therefore p \supset r$

Addition (Add):

p
$\therefore p \vee q$

Conjunction (Con):

p
q
$\therefore p \wedge q$

Conditional Argument (CA):

p Assume
...
q
$\therefore p \supset q$

reductio ad absurdum (RA):

$\neg p$ Assume
...
q
...
$\neg q$
$q \wedge \neg q$
$\therefore p$

PROVING THEOREMS

A theorem of a formal system is one that can be derived from the definitions, rules, and postulates of the system. Rather than being proved for particular assertions (represented by capital letters), these proofs generalize over all relations of assertions (or actions) (represented by lowercase letters). Consider the following cumbersome assertion: If, if all tigers are cats then all cats are mammals, then if, if all cats are mammals then all mammals are hairy, then if all tigers are cats then all mammals are hairy. Formally,

$$(T \supset M) \supset [(M \supset H) \supset (T \supset H)].$$

This is sometimes called Principle of the Syllogism and shows that a material implication (e.g., if T then M) implies a third or middle term related to each of the original terms. This is a theorem of the system at hand because it can be proven without
premises. The proof of The Principle of the Syllogism in its general form is as
follows:

$$\therefore (p \supset q) \supset ((q \supset r) \supset (p \supset r))$$

1. $p \supset q$	Assume
2. $q \supset r$	Assume
3. p	Assume
4. q	1, 3 MP
5. r	2, 4 MP
6. $p \supset r$	3-5, CA
7. $(q \supset r) \supset (p \supset r)$	2-6, CA
8. $(p \supset q) \supset ((q \supset r) \supset (p \supset r))$	1-7, CA

6.6   GRAPHICAL PROOFS OF VALIDITY

Proof of validity using postulates, equivalences, and valid argument forms,
often called the method of "natural deduction," generally follows a process
that is comparable to the one we might use in making an argument to someone
else. Even though the language becomes clumsy, one can usually convert a
deductive proof of validity into an argument in something like ordinary language. Consider the proof of the argument form transitivity. Let A mean "The
economy worsens," B mean "unemployment rises," and C mean "Consumer
sales decline." The argument would look this way formally:

1. $A \supset B$	Premise
2. $B \supset C$	Premise
3. A	Assume
4. B	1, 3 Modus Ponens
5. C	2, 4 Modus Ponens
6. $A \supset C$	3-5, Cond. Arg.

The formal description can also lead to a version of the argument in ordinary
language:

> If the economy worsens, then unemployment rises and if unemployment rises,
> then consumer sales decline. Suppose that the economy does worsen. Given the
> first premise, this means that unemployment rises. But if unemployment rises,

given our other premise, it will follow that consumer sales decline. So, given the premises, if the economy worsens, then consumer sales will decline.

The ordinary language argument follows (more or less) the formal argument.

An alternative proof procedure is based on the approach used in *reductio ad absurdum* arguments. Given certain premises, if we assume the negation of the conclusion and eventually arrive at a contradiction, then the assumed negated conclusion must be false and so we can assert its negation, that is, the conclusion that we sought to prove. Another way of achieving this same result is to connect the premises and assumed negated conclusion through a tree-like structure (sometimes called "truth-trees" and sometimes called "tableaux"). In general, a tree represents a series of conjunctions (marked by two propositions related vertically) and disjunctions (represented by branching divisions) that show in each case what must hold if a given proposition is true. Each of the operations that occur in the logic of material implication (and so not including the exclusive disjunction or strict implication which are modal operations) can be represented by a vertical connection or a branch. In order for an inclusive disjunction, $A \vee B$, to be true, then, at minimum, one of the disjuncts must be true. This can be represented in a tree where the disjunction leads to two branches to indicate that either A is true or B is true.

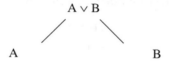

If the disjunction is negated, however, then the resulting tree is one that shows that A *and* B are false (and so ¬A and ¬B are true).

This vertical branch can also be represented without lines so that the following is equivalent to the branch above:

$$\neg(A \vee B)$$
$$\neg A$$
$$\neg B$$

Consider the tree for a conjunction, $C \wedge D$. In this case, the proposition is true if both C and D are true. This can be represented by a vertical relation that indicates that both terms are true in the same branch.

$$C \wedge D$$
$$C$$
$$D$$

If a conjunction is negated, then it is represented by a branching relation that shows that either one term or the other is false (that is, it shows that ¬C is true on one branch and ¬D is true on the other).

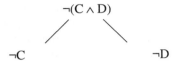

Since a material implication, E ⊃ F, is equivalent to an inclusive disjunction, ¬E ∨ F, implication can be represented by a branching relation indicating that either ¬E is true or F is true.

If the implication is negated, then it is equivalent to ¬(¬E ∨ F) and so can be represented by a vertical, non-branching relation.

$$\neg(E \supset F)$$
$$E$$
$$\neg F$$

Biconditionals, G ≡ H, for example, combine two conditionals, G ⊃ H and H ⊃ G and so the resulting tree shows that the biconditional is true when either G and H are true or G and H are false.

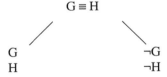

A negated biconditional can be represented as the following:

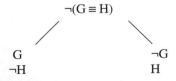

Finally, a double negation is true if and only if the original term is true. This can be represented as a vertical relation.

$$\neg\neg A$$
$$A$$

A branch closes when it contains propositions of the forms p and ¬p, indicating that the branch leads to a situation in which both p and ¬p must be taken. Since this state of affairs is not possible, it means the particular branch is not viable and may be closed (marked with an 'X').

In order to prove an argument valid, list the premises and the negated conclusion. Apply all of the tree rules that can be applied. Once a branch closes, then no further rules can be applied to that branch. If all of the branches close, then every possibility that follows from the original premises and the negated conclusion have led to contradictions. As a result, we can conclude that the original premises and the original conclusion are such that if the premises are true, the conclusion must also be true.

Consider again the proof of transitivity. Given the premises A ⊃ B and B ⊃ C, it follows that A ⊃ C.

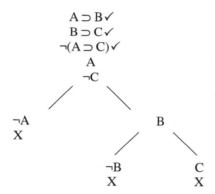

In this case, each of the branches closes and so we may conclude that the proposed argument is valid. *It is useful when applying the tree rules to check off each line as the relevant rule is applied and to apply those rules first that do not branch.*

PROVING THEOREMS

Theorems can also be proven using the method of tableaux. Begin by assuming the negation of the conclusion (the negation of the theorem to be proven) and apply the tree rules just discussed. If all of the branches close and since there are no premises, there is no case in which the negated theorem can be true and not lead to a situation in which incompatible alternatives must be taken. The negation of the theorem is therefore false and the theorem is true in every case. Here is a proof of the Principle of the Syllogism proven by natural deduction above.

$$\therefore (p \supset q) \supset ((q \supset r) \supset (p \supset r))$$

$$\neg[(p \supset q) \supset ((q \supset r) \supset (p \supset r))] \checkmark$$
$$p \supset q \checkmark$$
$$\neg((q \supset r) \supset (p \supset r)) \checkmark$$
$$q \supset r \checkmark$$
$$\neg(p \supset r) \checkmark$$
$$p$$
$$\neg r$$

```
               ¬q                    r
                                     X
           ¬p        q
            X        X
```

If an argument is invalid, then there is at least one assignment of truth (true or false, taken or failed) that can show that there is a situation in which the premises are true and the conclusion is false. This is especially easy to find using trees. If an argument is invalid, there will be at least one branch of the tree that does not close. The values of the terms can be determined by simply reading the values of each term as expressed on the open branch. Consider the following argument:

$$p \supset (q \wedge r), \neg r \therefore \neg p$$

$$p \supset (q \wedge r) \checkmark$$
$$\mathbf{\neg r}$$
$$\neg\neg p \checkmark$$
$$\mathbf{p}$$

```
        ¬p                        q ∧ r ✓
         X
                           q              r
                                          X
```

In this case, the left-most and right-most branches close. An assignment of values that demonstrates that the original argument is invalid can be read from the middle branch. In this case, when q is true, p is true and r is false, the premises are true and the conclusion false.

6.7   EXERCISES

6.7.1. Symbolize each argument using the letters suggested. Construct formal deductions for the valid arguments below. Each of these can also be proven using the method of truth-trees.

1. If Darnell is lost, then Frank will find him. Darnell is lost and Edward is just arriving. So, Frank will find Darnell. (D: Darnell is lost. F: Frank will find Darnell. E: Edward is arriving.)

2. If it is not the case that if all Republicans are conservative, then the Democrats will lose, then a better disaster plan can be developed. A better disaster plan cannot be developed and the Democrats will not lose. So, it is not the case that all Republicans are conservative. (R: All Republicans are conservative. D: The Democrats will lose. B: A better disaster plan can be developed.)

3. If the president of the university meets with the activists and he accepts their demands, then the diversity plan will be instituted. The president, in fact, accepts the demands. If the diversity plan is accepted, then both resources will be available and there will be no more protests. The president meets with the activists. Therefore, there will be no more protests. (P: The president meets with the activists. A: The president accepts the demands. D: The diversity plan will be accepted. R: Resources will be available. N: There will be protests.)

4. Everyone can tell right from wrong. But if moral judgment is a personal matter, then there is no moral law. Of course, there is a moral law, if J. S. Mill was correct. If everyone can tell right from wrong, then Mill was correct. Therefore, moral judgment is not a personal matter. (E: Everyone can tell right from wrong. P: Moral judgment is a personal matter. L: There is a moral law. M: Mill was correct.)

5. Cindy can go to the coast this weekend or she can stay at home; she cannot do both. If she wants to study, then she will need to stay home. If she wants to pass the class, then she wants to study. She definitely wants to pass the class. Therefore, she cannot go to the coast. (C: Cindy can go to the coast this weekend. H: Cindy can stay home this weekend. P: Cindy wants to pass. S: Cindy wants to study.)

6.7.2. Prove the following problems for practice. Here the uppercase letters indicate particular assertions that represent arguments that could be given in ordinary language. Rather than "translate" these arguments from ordinary language to start, we will simply begin with them represented in standard form.

1. $A \mid B \therefore \neg B \vee \neg A$

2. $C \mid D, C \therefore \neg D \vee E$

3. $(A \supset B) \mid C, C \therefore A$

4. $(F \supset G) \mid (\neg G \supset C) \therefore F \supset \neg C$

5. $\neg(A \supset B) \therefore A$

6. $\neg D \therefore C \supset \neg D$

7. $A \supset (B \wedge C), \neg B \therefore \neg A$

8. $(\neg F \supset \neg G) \supset \neg L \therefore L \supset G$

9. $\therefore (G \supset H) \vee (H \supset G)$

10. $\therefore [A \vee (\neg A \wedge C)] \equiv (A \vee C)$

11. $\neg(A \supset B) \supset C, \neg C \wedge \neg B \therefore \neg A$

12. A ⊃ (Z ∧ Z), ¬A ⊃ Z ∴Z

13. A ⊃ B, C ⊃ D, (B ∨ D) ⊃ E, ¬E ∴¬(A ∨ C)

14. C ⊃ (D ∨ ¬(A ∨ B)), ¬A ⊃ B ∴¬D ⊃ ¬C

15. D ∨ ¬A, ¬(A ∧ ¬B) ⊃ ¬C, ¬D ∴¬C

16. A ∧ (B ⊃ C), ¬(C ∧ A) ∴¬B

17. ¬A, (A ∨ B) ≡ C, ¬B ∴¬(C ∧ D)

18. D ⊃ B, D ⊃ (B ⊃ W), B ⊃ (W ⊃ S) ∴D ⊃ S

19. A ∨ B, C, (A ∧ C) ⊃ D ∴D ∨ B

20. J ⊃ R, (¬R ⊃ S) ⊃ D, ¬D ∴¬J

6.7.3. Use the method of tableaux to prove 6.7.2 problems 5 through 20.

6.7.4. The following arguments are not valid. Use the method of tableaux to determine an interpretation of the terms where the premises are true and the conclusion false.

1. E ⊃ (G ∧ H), G ∴¬F

2. K ⊃ L, M ⊃ N, (L ∨ N) ⊃ O, O ∴¬(K ∨ M)

3. ∴[(A ⊃ B) ⊃ B] ⊃ B

4. D, E ⊃ F, ¬F ∨ ¬D ∴E

5. A ∧ (¬C ∨ D), ¬(B ⊃ D) ∴C

6.7.5. The following questions focus on the modal aspects of the exclusive disjunction. Except where indicated, answer each question using an "informal proof," that is, an explanation that relies on the concepts of the system, but without giving a line-by-line demonstration.

1. Does the following form of entailment hold?

$$p \Rightarrow p \mid q$$

2. Suppose it is the case that □¬(p ∧ q). Can ◇(p ⊃ q) also hold?

3. If □(p ∧ q), can ¬◇(p ∧ ¬q) also hold?

4. If p ∣ (q ∧ t) and ¬q is equivalent to r and r ⊃ (s ∨ t), then can we conclude that p ⊃ s?

5. If R ∣ M and ¬◇(S ∧ ¬M), then does □(R ∧ S) also hold?

6. If p ∣ q and q = r and r = s, does ¬p = s also hold?

7. Consider the following proposed form of entailment:

$$[(p \supset \neg q) \wedge p] \Rightarrow p \mid q$$

Explain why premises of the form stated do not entail the exclusive disjunction shown.

8. Suppose p ∣ [(q ⊃ r) ∨ (t ⊃ s)]. Prove that p ⊃ ¬s. Use a line-by-line demonstration.

CHAPTER *7*

AN OVERVIEW OF QUANTIFIED LOGIC

7.1   INTRODUCTION

The proposition p | q represents a relation between two propositions or actions in which at least one holds (or is true) and at most one holds (or is true). Propositions p and q are complete statements in that they include a subject and a predicate. Of course, the lower case 'p' and 'q' represent any complete assertion at all and the compound assertion, p | q, is the exclusive disjunction of two propositions, which will be true just in case the propositions represented are incompatible, that is, they exclude each other. For example, 'p' might be replaced by the assertion "I will go to the beach tomorrow" and 'q' by the assertion "I will work in my office all day." In this case, p | q is true because the propositions expressed by p and q are, in fact, incompatible—I must carry out one or the other, but not both.

We can represent particular propositions of this sort using upper case letters and so represent arguments of the following sort.

> If Abe has studied Euclidean geometry, then he knows how to reason deductively.

Abe has studied Euclidean geometry.

So, Abe knows how to reason deductively.

Here, let 'G' stand for "Abe has studied Euclidean geometry" and 'D' stand for "Abe knows how to reason deductively." In this case, we can represent the argument this way:

$$G \supset D$$
$$G$$
$$\therefore D$$

The *logic of assertions* of the sort we discussed in the last chapter provides a means to prove the validity of arguments composed of complete propositions and compound propositions (statements that include at least one complete proposition and one or more logical operators). These proofs of validity use rules and postulates to construct an argument connecting premises and conclusion by means of a finite number of valid propositions. *Equivalence rules* allow the conversion of propositions into logically equivalent expressions, postulates of the system show the structure of entailment and allow the assertion of what follows necessarily from given premises, and *truth functional arguments* allow one to use already-made assertions as premises for new valid conclusions.

The logic of assertions, however, is limited in that it does not allow for the differences that make the arguments of standard syllogisms valid. Consider

All humans are mortal.

Joe Biden is human.

So, Joe Biden is mortal.

Notice that each proposition is a complete assertion, subject and predicate. Let 'H' stand for "All humans are mortal", 'J' for "Joe Biden is human", and 'C' for "Joe Biden is mortal." In this case, each uppercase letter represents a complete assertion. We can then represent the argument as follows using assertions:

$$H$$
$$J$$
$$\therefore C$$

Despite the fact that we have proved an argument of the sort given above to be a valid form using Venn diagrams and axioms, the logic of assertions does not provide us any means of proving the validity of the argument. In this case, the important relations are masked by the analysis that stops at the level of complete assertions.

In order to represent subjects and predicates we need a more complex system that takes into account the important features of categorical propositions: quantity, quality, and distribution. By representing all complete statements as the same (any assertion can be used to replace 'p' in our formulas), the distinctive character of categorical statements is removed from consideration and validity of a syllogism cannot be proved.

Consider the claim "Emma is a human." The claim marks the expectation that there is a mode of action that we call "being human" and that Emma, a particular individual, is a human or, in terms of modes of action, interacts with the world in a human way. In our discussion of the syllogism, categories such as human being and animal are understood as collections of particulars that share a common quality (or a common mode of action). In order to talk about the particulars, Socrates or Emma, we are required to treat individuals as complete categories of their own. Since categories (and individuals) are a product of both what things present (i.e., how they act) and the interests and actions of others (i.e., processes of selection), to say "Emma is a human being" is to put one category in relation to another, to put Emma in relation to others in virtue of a mode of action and set of interests. Another way to represent this relation is to label the category 'human' and label the individual, in this case Emma, 'e'. The relation can then be represented symbolically 'He', or "Emma is human."

It is easy to understand the relation expressed "He" as placing an individual in relation to a class of things. It is useful to keep in mind that the individuals themselves are also categories (or things actively individuated) and so are not radically different in kind from the categories to which they are related (e.g., human, animal, good, acerbic). Nevertheless, the particulars in this system are taken to be "atomic" in the sense that they seem to have been individuated in advance of their relation to the category in question. We say that Emma (presumably a being already) is a human (is taken in relation to this larger category). At the same time, we recognize that Emma *is* in part thanks to the relation she already holds to the category of being human. In order to understand the ordering character of relations, however, we simply treat individuals as if they can be brought into a relation or discovered to have a relation with something else without, at the moment, considering "where" the individuals came from. Once again, ontology and epistemology are bound together in the process of ordering.

PERSONS IN PLACE[1]

In their co-authored work, *Power and Place: Indian Education in America*, Vine Deloria, Jr. and Daniel Wildcat offer a conception of individuals and relations that bridge epistemonlogy and ontology. The philosophical starting point for *Power and Place* is "the realization that the world, and all its possible experiences, [constitutes] a social reality, a fabric of life in which everything [has] the possibility of intimate knowing relationships because, ultimately, everything [is] related" (Deloria and Wildcat, 2001, 2). According to this view, the relation of things is grounded in two

---

[1] This section is taken from an earlier paper (Pratt, 2006, 6–7).

"experiential dimensions": power and place. The first of these, power, is at once a unifying force and, at the same time, an individuating one.

The notion of power in native traditions has often been discussed. In an early formulation of the idea for a Western audience, William Jones, a Mesquakie Indian and a student of Franz Boas at Columbia University, argued that the notion of power, here called "*manitou*," is a notion held by traditional Algonquian people that marks both what is common among things and what makes things different from one another. For Jones, the claim that Manitou is both commonality and difference also shows the notion to be primitive and contradictory. Jones observes that Manitou is "an impersonal essence" (1905, 185) that exists "everywhere in nature" (1905, 190), while at the same time it marks "the quality of self-dependence" and "rests on the perception of a definite, localized personality" (1905, 184).

Writing a few years earlier, Tuscarora ethnologist J. N. B. Hewitt identifies a comparable notion in Iroquoian thought which he calls "*orenda*" (1902). For him, the contradictory character marks the "primitive" and "barbarous" nature of Native thought. Despite Jones' worry about the contradictory character of the ideas and Hewitt's dismissal of it, the idea of *manitou* or *orenda* nevertheless provides a ground for fostering diverse personalities even as it makes understanding and unification an ongoing possibility.

In a way, the notion of power is like the notion of continuity. While continuity marks the connections between things it also implies the possibility (and serves as the medium) in which differentiation occurs. At the same time, power seems to be more than continuity alone in that it also marks a motive force that gives direction to the process of differentiation. As Wildcat observes, power "moves us as human beings" and "inhabits or composes" "all of the connections or relations that form the immediate environment" (Deloria and Wildcat, 2001, 140). From this perspective, power is its own activity, the exercise of power, and the bonds or relations that constrain and enable that activity. At the same time, in action, power also takes on the character of particular beings so that relations are interactive with other agents, not self-action alone.

The notion of *orenda* helps to illustrate the implications of power. Like *manitou*, *orenda* is understood as a kind of unifying notion in that everyone has *orenda*, but it is also a differentiating notion in that different people have different *orendas*. One way to understand this concept is suggested by the linguistic root for *orenda*, -*ren*-, which is also the root for the terms for "song," "to sing," and "voice" or "speech." From this angle, *orenda* marks the song or voice of particular things. While my voice and yours are distinctively our own, they also are similar in important ways. One can distinguish between individuals on the basis of their characteristic song or voice, and one can also link individuals together through similarities of voice—much as one might link the sounds of several drums, where each makes its own sound, but together the sounds are like enough that they literally resonate. Listening to the sound together, the resonating drums are a unity of sound; listening for the characteristic expression of a single drum makes individuals emerge from the collective sound. It is important to note that while each drum has its own sound, it is both a sound dependent upon the drum's origin (the skins and wood of which it is made) and its interaction with other agents—the drummer, the listeners, even the other drums in its hearing. In this sense, the distinctive character or power of a given individual is also dependent upon a host of relations and other persons. The bond with others is an indissoluble one in that the characteristic expression of one person is what it is only in a complex of relations. This complex is called place.

The choice of "place" as the term to characterize the relations that frame power is important. "Place," as Wildcat observes, "is not merely the relationship of things, resources, or objects, it is the site where dynamic processes of interaction occur— where processes between other living beings or other-than-human persons occur" (Deloria and Wildcat, 2001, 144). On this view, a person *is* how she or he acts or rather interacts. The point is that there are, in this view, no "inner" or "real" essences that determine what a thing is. Just as the distinctive voice of a particular thing is an interactive process among things, so what a thing is cannot be either an essence in isolation or a particular in relation to an abstract category.

The result of this view is that, on one hand, individuals are instances of kinds, that is, their characteristic expressions are similar to the expressions of others. A human being is a human being in that she or he has characteristic expressions like those of other humans and not like those of bears or fish. To the degree that a person has the expressions characteristic of a bear or a fish, she or he is literally partly bear or fish. On the other hand, individuals are truly individual in the ways in which their characteristic expressions are unlike those of others or the ways in which their combination of characteristics is distinctive. From this perspective, who one is becomes a matter both of similarities in the exercise of power and the ways in which one's power is not just human or bear or fish. A person is who she is neither in isolation nor in terms of abstract relations and so, in this sense, a person *is* only in place.

"Place" is also an important term in that it implies the importance of land among the others who form the context of a person. From one angle, land and sustaining environments are simply a necessary condition for the existence of any particular beings on earth. From another angle, land, rivers, ecosystems, even farm fields, hills, and mountain ranges are also persons in their own right in interaction with other persons, human and otherwise. In a universe of interaction, a particular land is understood to be another characteristic expression of power, different from the expressions of animals or plants, so an agent in its own right.

A further implication of this view is that persons are not *only* persons in their own right. A person, in Deloria's sense, can at once be an individual and a part of a "larger" agent such as tribe or nation. One can be a human person and part of the personality of an ecosystem. As a result, Deloria offers us a world that is not compositional in the usual sense. On many traditional Western accounts (most forms of realism, for example), the universe is made of atoms of one sort or another that, taken together, form the macrocosm that we experience. Deloria's universe does not rule out a macrocosm composed of "smaller" elements, but there are no smallest units or largest ones. As a result, each person is always both a person in his or her own right and bound up in the personality of something else. Such relations are not fixed and unchangeable but emerge and disappear as their relations (and so their personality) changes. Wildcat identifies this notion of emergence as "a model of change or development where change is not reducible to a discrete factor or factors, but rather the interaction of multiple factors or causes that are understood as processual in character as opposed to mechanical" (Deloria and Wildcat, 2001, 15).

If, as Deloria argues, the universe is made up of persons, and if interactions among persons are moral relations, then all relations in the universe have a moral character. When interactions are considered morally, the operative standard Deloria and Wildcat each argue for is one drawn from the nature of personhood itself. "There is," Deloria concludes, "a proper way to live in the universe: There is a content to every action, behavior, and belief. ... There is a direction to the universe, empirically exemplified in the physical growth cycles of childhood, youth, and old age, with the corresponding responsibility of every entity to enjoy life, fulfill itself, and increase in wisdom and the spiritual development of personality" (Deloria, 1999, 46). Put another way, actions and agents that promote growth and development are good and those that do not are bad. The standard is clearly a relative one in that actions will be evaluated according to their potential to promote growth in terms of relations in which they stand, that is, in terms of their places. Since persons, in this model, include human beings, individuals other than human beings, as well as unified systems like tribes and ecosystems, moral relations occur at every scale and evaluations of actions and agents will occur at every scale as well.

Just as Royce argued that logic provided a key starting point for the study of science, social theory and ethics, Deloria provides a means to see how the logic of relations is bound at once to ongoing interactions with others, to a conception of knowledge framed by the activities that emerge in those interactions, and a conception of moral theory.

7.2 REPRESENTING RELATIONS

The claims that Emma is a human or a mother or an African American or a corporate president mark relations from one angle and modes of action from another. When individuals are "assigned" to predicates (e.g., He) in the context of logic, there is a risk that an ontology is also implied that recalls those at issue in the problems of logic and in the tension between realism and nominalism. On the contrary, "assignments" in logic occur in a context of inquiry where one is trying to understand a particular system of order or considering possible courses of action. The ontology this perspective implies is one that calls into question the idea that the "assigned" relations are given in advance or are independent of the situation at hand. Instead, the emergence of individuals and relations is a process of active ordering by agents who gain power and purpose through their ability to be agents even as they are constrained by the agency of others and the lifeworld in which they live.

The logic of propositions masks the process of relating predicates and individuals because the relations are given within the claims made. That Emma is an excellent student already takes Emma as related to the categories of excellence and study. Predicate logic makes the association apparent by explicitly representing the subject and predicate but in so doing also reinforces the idea that Emma is an individual potentially separate from every predicate except identity (if we take identity as a predicate) and that predicates are things that can be added or subtracted from such individuals. From the conception of logic developed by Peirce, Royce, and Dewey this notion of given individuals and predicates is not foundational but is rather the consequence of foundational acts of judgment past. These acts that once brought a certain order to the world of experience became habits of thought and institutions that helped to maintain the established systems of order. What is at stake in the study of logic is the ability to examine the structure of the foundational judgments in a way that can add to an understanding of the systems of order that structure lifeworlds and provide an opening for criticism and transformation of those systems.

We can consider the relation of predicates and individuals by representing the relations internal to a proposition using the following resources.

> *Let uppercase letters A, B, C, … abbreviate modes of action* (or what we have called predicates).
>
> *Let lowercase letters a, b, c, … abbreviate proper names* (i.e., lowercase letters stand for individuals or the results of the process of individuation).

So if 'e' stands for Emma and 'H' stands for the mode of action, "being human" or "acting human" or simply "is human," "Emma is human" is represented:

$$He$$

And if 'S' stands for the mode of action of being a student and 'W' for being a woman, then we can represent "Emma is a student" and "Emma is a woman" this way:

$$Se$$
$$We$$

And we can represent "Emma is a human, a student, and a woman" in this way:

$$He \land Se \land We$$

Now, when we suppose that we are discussing some randomly selected human (perhaps Emma, but it could be anyone), we can replace the proper name with a *variable*.

> *Lowercase letters u, v, w, x, y, and z all serve as variables* and stand for some individual not yet identified.

If we simply replace Emma with a variable we get Hx, which can be read as "x is a human" or "x acts humanly." This expression, however, is incomplete because it does not refer to anything at all—that is, the x is ambiguous in its reference. Does it refer to one particular individual or all of them? In fact, a variable attached to a mode of action in this way is called an *open variable* and the expression it is part of is called an *open formula*. In order to make a complete proposition, the variable must be *bound*, that is, given some referent, even if it is not to a particular individual.

There are three ways to make the open formula Hx a complete proposition, that is, three ways to bind the variable, x:

1. Change the 'x' to a name that refers to an individual (e.g., 'a' for Adam, 'e' for Emma).
2. Make the 'x' refer to some particular individual that is a human. This is equivalent to the claim "Someone is a human" or "There is an individual who is a human."
3. Make the 'x' refer to everything so that anything you pick out will complete the sentence. This is equivalent to "Everything is human."

Of course, following Habermas and Dewey among others, no proposition has meaning without a context in which it can stand in relation to individuals, modes of action, and other statements. The context in which a proposition has meaning is the *universe of discourse* and *unless it is explicitly narrowed, the universe of discourse includes everything that has been individuated.*

So if the expression Hx has meaning, it will be against a background or universe of discourse that includes all individuals identifiable in the present context or lifeworld. One of those things is, of course, Emma and we can say, He (or Emma is a human). This is the first way to bind a variable—replace it with the name of a particular individual.

The second way to bind the variable follows from the first. If Emma is a human, then it follows that there is, in general, at least one thing that is human. We will symbolize this statement

$$\exists x Hx$$

that is, "There exists an x such that x is H," and call the result an *existential generalization*. Notice that the variable, x, is now bound because the initial symbol '∃x' means "There is something, x." In effect, the variable has been made part of a claim that asserts that there is something that has the mode of action H in the universe of discourse.

The third way to bind the variable in the statement Hx is make it refer to all of the individuals in the universe of discourse. Notice that this is comparable to the idea of a distributed term in the theory of the syllogism: the expression 'Hx' is then taken to be the claim that everything there is has the mode of action H. In this case we say "Everything is human" or "For all things, they are human." This is called a *universal generalization* and is represented

$$\forall x Hx$$

Of course, given that we can fill in x with anything at all that is individuated (including the Lincoln Memorial or Grant Wood's *American Gothic*) it is clear that the statement is false—not everything acts "humanly." Yet, regardless of whether the statement is true or false, the symbol '∀x' will be taken to mean that the mode of action (or actions) under its "control" or "scope" applies to every individual in the universe of discourse.

What has been said about the expression Hx can be generalized to apply to any mode of action. We can represent this more generally by 'fx' where *the lowercase f, g, h, … represents a variable to be filled in by a particular mode of action* (say "acting human"). Given this generalization, we can summarize the above discussion by saying that an open formula, fx, can be completed in three ways:

1. *Singular Sentences* where 'x' is replaced by the name of some individual, for example, 'fa' or 'fb'.

2. *Existential Generalization* where 'x' is replaced by some unspecified individual and is represented '∃xfx'.

3. *Universal Generalization* where 'x' is replaced by any individual available, represented '∀xfx'.

Although we will be concerned here only with monadic quantification, that is, the relation between a predicate or category and an individual, this same approach can be used to represent relations that involve more than one term. For example, if Tom owes Sue money, we can represent the relation in this way: Mts. Here, 'M' represents the two place relation 'x owes money to y', 't' represents Tom, and 's' represents Sue. The expression, 'Mts', then, can be read "Tom owes money to Sue." This approach can be used to represent relations of any number of terms.

7.3 THE MEANING OF QUANTIFIERS

To state the meaning of quantified sentences can be somewhat difficult, especially in cases where negation is involved. The following statements give an idea of how to interpret in English each of the formalized expressions. Here, 'B' will represent the expression 'is beautiful' or rather the mode of action of being beautiful.

∀xBx	Everything is beautiful.
¬∀xBx	Not everything is beautiful.
∀x¬Bx	Everything is not beautiful.
	or Nothing is beautiful.
¬∀x¬Bx	It is not the case that everything is not beautiful.
∃xBx	Something is beautiful.
¬∃xBx	It is not the case that something is beautiful.
∃x¬Bx	Something is not beautiful.
¬∃x¬Bx	It is not the case that something is not beautiful.

As should be clear, the statements above all provide ways of representing different quantities and qualities that will eventually permit us to represent categorical propositions in a way that will integrate them into the system of derivation rules.

Now, consider the assertion ∃xBx. If there is something in the universe of discourse that is beautiful, then it follows that it is not the case that all of the things in the universe are not beautiful, that is, ¬∀x¬Bx. This relation between ∃xBx and ¬∀x¬Bx or more generally between statements of the form "∃xfx" and "¬∀x¬fx" is an equivalence relation. There are four such relations among the various statements listed above. Since, in each case, the pairs are equivalent, they are like the equivalences of assertions that we discussed earlier. Should one appear on a line in a proof, the equivalent claim may also be validly asserted. The following formulas are equivalent:

$$\exists xfx = \neg\forall x\neg fx$$
$$\forall xfx = \neg\exists x\neg fx$$
$$\neg\forall fx = \exists x\neg fx$$
$$\neg\exists xfx = \forall x\neg fx$$

These equivalences can be summarized as two *Rules of Quantificational Equivalence* (abbreviated QE):

1. Two propositions are equivalent if they are exactly alike except where one has a negation sign followed by a universal quantifier, the other has an existential quantifier followed by a negation sign. For example, ¬∀xfx is equivalent to ∃x¬fx.

2. Two propositions are equivalent if they are exactly alike except where one has a negation sign followed by an existential quantifier, the other has a universal quantifier followed by a negation. For example, ¬∃xfx is equivalent to ∀x¬fx.

Now, consider the following proposition:

Emma is either a human or not.

This can be easily represented in the present symbolic system as He | ¬He, where 'H' is 'is a human' and '¬H' is 'not a human'. But if this is true, then it is also true that someone is either a human or not. This is represented

$$\exists x(Hx|\neg Hx)$$

Notice that *this statement is not equivalent* to another that might seem like a viable alternative: ∃xHx | ∃x¬Hx. This second assertion is read 'There is something that is human or (in the exclusive sense) there is something (potentially something else) that is not a human'. This second proposition entails the material implication, "if there is something that is human then there is nothing that is not human"—clearly not equivalent to "There is something that is human or not human."

In the statement, ∃x(Hx|¬Hx), the 'x's that appear attached to 'H' and '¬H' are both under the "control" of the existential quantifier, that is, the 'x's are said to be *under the scope* of the existential quantifier. In this case, the thing picked out by x is the same in both occurrences. In the alternate example, ∃xHx | ∃x¬Hx, there is no requirement that the same thing be picked out in each use of the 'x' because the 'x' in 'Hx' and the 'x' in '¬Hx' are not "bound" together under the scope of a single quantifier. As a result the second quantifier must be treated separately and so is equivalent (by the rules above) to ¬∀xHx, that is, "not everything is H." If ∃xHx | ∃x¬Hx is true and ∃xHx is true then ¬∃x¬Hx follows, which is equivalent to ¬¬∀xHx and so to ∀xHx or "everything is H."

7.4 EXERCISES

Using the predicates listed for each assertion, "translate" each of the following into quantified logical form.

1. No managers are sympathetic. (Mx, Sx)

2. Everything is in its right place. (Rx)

3. Some cell phones have no service here. (Cx, Sx)

4. Not everything is settled. (Sx)

5. *Radiohead* concerts are amazing. (Rx, Ax)

6. Nothing is everlasting. (Ex)

7. Not every earthquake is destructive. (Ex, Dx)

8. Very few people do not like Mac computers. (Px, Mx)

9. Only registered voters can vote in the next election. (Rx, Vx)

10. Not everyone disapproves (i.e., does not approve) of the president's cabinet selections. (Ax)

11. Grades are not always accurate. (Gx, Ax)

12. There is nothing that does not have an underlying cause. (Cx)

13. A person can be sued for libel if and only if they have written something against someone's character. (Px, Sx, Wx)

14. No one argues that the American economy was in good shape at the end of 2008. (Ax)

15. Some hybrid cars are bad for the environment. (Hx, Bx)

16. If all employees approve the work stoppage and went out on strike, they are prepared to go on unemployment. (Ex, Ax, Wx, Px)

17. Most people feel more could have been done to prevent the genocide in Rwanda. (Px, Fx)

18. Some party leaders have proven to be racist and sexist. (Px, Rx, Sx)

19. If a soldier is not asked, then that soldier is not required to tell. (Sx, Ax, Tx)

20. If some party leaders have proven to be racist and sexist, then these people should be voted out. (Px, Rx, Sx, Vx)

7.5 RULES OF QUANTIFICATIONAL LOGIC

It is clear that if there is an individual such as Emma, who is either a human or not, we can generalize the fact by claiming that there is something that is either human or not. If we want to go the other way, given that something is either human or not, we may *instantiate the variable* by picking out an individual that satisfies the conditions of the modes of action. In this case, Emma satisfies those conditions so we can instantiate the variables with the abbreviation of Emma's name, e. The result is that we can write 'He | ¬He'. This procedure can be followed any time we encounter an existentially quantified proposition. In effect, if the statement is true, then there is at least one unspecified individual that fits the requirements of the relations given. The result is a new rule of inference:

> *Existential Instantiation* (EI). From any existentially quantified proposition (any proposition whose main symbol is the existential quantifier), we can infer *any instance of it, provided that the individual introduced has not yet appeared in the deduction.*

The qualification that an existential instantiation must use an individual not already mentioned in the deduction is to guard against the following. Suppose that we instantiate ∃xSx with the individual named Adam, a, where S is the mode of action of being a student. Now, some things in the universe of discourse are students and other things are not. So it should follow that there is something that we can find that we can use to fill in the variable in ∃x¬Sx, that is, there is something that is not a student. If, however, we instantiate ∃xSx with Adam, a, and then instantiate ∃x¬Sx with the same individual then we get

$$Sa \land \neg Sa$$

To avoid the obvious contradiction resulting from two statements that are not necessarily contradictory, *always instantiate an existential generalization with an individual who has not already appeared in the deduction.* When we do that in the sample we get

$$Sa \land \neg Sb$$

This result is not necessarily a contradiction and, if it is, it will be due to other factors (e.g., size of the universe of discourse or the assignment of names or other statements that are also true).

Now consider the claim "Everything is either beautiful or not beautiful." This can be symbolized

$$\forall x (Bx | \neg Bx)$$

Suppose that we would like to consider what this implies for some particular individual in the universe of discourse, perhaps Adam, a. We should be able to use any individual at all to replace the variable in the given statement and the statement should remain true. To do this, we need another rule of inference.

> *Universal Instantiation* (UI). From any universal generalization (any assertion whose main symbol is the universal quantifier) we may validly infer *any instance* of it.

Since the meaning of the universal quantifier is everything or for everything in the universe of discourse, it follows that we can introduce any particular instance. So, we can choose to replace the variable with Adam in the above statement and so get

$$Ba|\neg Ba$$

Notice that if we have two premises, one existential and one universal, if we instantiate the universal generalization first, then we must instantiate the existential generalization with a *different* individual. Consider this simple deduction.

1. $\exists xBx \supset T$ Premise
2. $\forall xBx$ Premise $\therefore$ T
3. Ba   2, UI
4. Bb $\supset$ T   1, EI

Because these terms are instantiated with different individuals and, as a result, the proposition on line 3 does not match the antecedent of the material implication on line 4, we cannot use the truth functional argument, *modus ponens*. We therefore cannot prove that the conclusion, T, follows validly from the premises. If, however, we instantiate the existential generalization first, then we can instantiate the universal generalization using the same individual (following the EI and UI rules) and so get

1. $\exists xBx \supset T$ Premise
2. $\forall xBx$ Premise $\therefore$ T
3. Ba $\supset$ T   1, EI
4. Ba   2, UI
5. T   3, 4 MP

Always instantiate existential generalizations first in order to have statements that allow for the use of equivalences and truth-functional arguments.

While the instantiation postulates allow one to select an individual with which to test a proposition that is quantified (one that concerns "all" or "some"), there should be a way to move from an individual case to a larger generalization. For example, if it is true that Emma is a human, then we should be able to validly infer that *someone* is a human. This can be concluded by using the following rule:

> *Existential Generalization* (EG). $\exists xfx$ can be inferred from any proposition fx where 'x' stands for an individual. (Note: 'fx' stands for some formula containing one or more x's.)

So, for example, He implies ∃xHx. Suppose the following argument: there is something where, if that thing is subject to gravity, then that thing will fall toward the center of gravity. Everything is subject to gravity. So, there is something that will fall toward the center of gravity.

1. ∃x(Gx ⊃ Fx)	Premise	
2. ∀xGx	Premise	∴ (∃x)Fx
3. Ga ⊃ Fa	1 EI	
4. Ga	2 UI	
5. Fa	3, 4 MP	
6. ∃xFx	5 EG	

In effect, since the argument proves that the thing, a, is a thing that will tend to fall toward the center of gravity, we can conclude that there is *something* that will tend to fall given the conditions stated in the premises.

Under certain conditions, one can also generalize universally. When, in a proof, you use a "randomly" selected individual (i.e., one that does not appear anywhere earlier in the proof), you can use the postulate UI to generalize across everything in the universe of discourse.

> *Universal Generalization* (UG). ∀xfx can be inferred from any proposition fx where 'x' stands for an individual that does not appear in the premises or as a result of rule EI. (Again note: 'fx' stands for some formula containing one or more x's.)

So given the assertion Ra in a proof where "a" was introduced by the postulate UI and appears nowhere earlier in the proof, you can conclude that ∀xRx.

Consider the argument: Everything has mass. So, for everything there is, if there is something with weight, there is something with mass.

1. ∀xMx	Premise	∴ ∀x(Wx ⊃ Mx)
2. Ma	1 UI	
3. Ma ∨ ¬Wa	2 Add	
4. ¬Wa ∨ Ma	3 C	
5. Wa ⊃ Ma	4 DI	
6. ∀x(Wx ⊃ Mx)	7 UG	

Here, a is introduced through UI. The assertion ¬Wa is introduced after the use of UI and uses the truth functional argument addition. Since "a" fits the criteria specified by UG, the conclusion follows using Universal Generalization.

7.6 EXERCISES

Translate each of the following arguments into quantified form and prove that each is valid using natural deduction. The letters that follow each argument give the predicate letters to use in symbolizing the argument.

1. All carpenters are precise and some handypersons are carpenters. Therefore, some handypersons are precise. (C, P, H)

2. All lawyers who did not take a law degree are incompetent. However, it is not the case that some lawyers are incompetent. Hence, all lawyers took law degrees. (L, D, I)

3. All activists are loyal to a cause and all people in power conserve the status quo. Since no one who is loyal to a cause will conserve the status quo, it is not the case that some activists are people in power. (A, L, P, C)

4. Either some farmworkers will attend or some mediators will attend. If anyone attends, then some of the mediators are both clever and diplomatic. So, some diplomats are clever. (F, A, M, C, D)

5. If all store supervisors are wise, then some employees benefit. If there are some store supervisors who are not wise, then some employees benefit. As you can see, either way, some employees benefit. (S, W, E, B)

6. If someone studies philosophy, then all students benefit. If someone studies literature, then there are some students. So if someone studies philosophy and literature, then someone benefits. (P, B, L, S)

7. Everyone is a Democrat or a Republican, but not both. If someone is a Democrat, then she is a liberal or a conservative. All conservatives are Republican. So all Democrats are liberal. (D, R, L, C)

8. If there are any mavericks, then all politicians are committed to change. If there are any politicians, then anyone who is committed to change is pandering. So, if there are any mavericks, politicians are pandering. (M, P, C, A (for "pandering"))

9. If everyone is a liberal, then no one is a conservative. There is a governor of Alaska and she is a conservative. So at least someone is not a liberal. (L, C, G)

10. If there are any cherries, then the raspberries are tasty, if they are sweet. Some cherries are ripe and some raspberries are sweet. Therefore, some raspberries are tasty. (C, R, T, S)

7.7 THE VALIDITY OF SYLLOGISMS

It is now possible to represent and prove the valid forms of the syllogism in Aristotelian logic. First, recall the standard form categorical assertions.

A—All humans are mortal.

E—No tigers are dogs.

I—Some students are juniors.

O—Some students are not juniors.

Let 'H' stand for the mode of action of being a human and 'M' stand for the mode of action of mortality. Recall the Venn diagram of A assertions.

In describing it, we say that if a thing is in the subject category, then it is also in the predicate category (because *all* S are P). This description suggests that the way to understand the relation expressed in A assertions is in terms of a conditional proposition. Using the symbolization we have discussed, we can represent the *A assertions* this way:

$$\forall x\,(Hx \supset Mx)$$

This can be read, "For all things in the universe of discourse, if a thing is human, then it is mortal" or "For all x, if x is human then x is mortal." In effect, if a thing, x, is in the human region of the Venn diagram of these terms, then it is also in the mortal region.

The *E assertion*, "No tigers are dogs," can be represented as follows, where 'T' stands for 'is a tiger' and 'D' stands for 'is a dog'.

$$\forall x\,(Tx \supset \neg Dx)$$

This case seems straightforward: for all x, if x is a tiger then x is not a dog.

We can represent the given *I assertion* this way:

$$\exists x\,(Sx \wedge Jx)$$

or "there is something that is both a student and a junior." The *O assertion* is also represented using an existential quantifier:

$$\exists x\,(Sx \wedge \neg Jx)$$

or "there is something, x, that is both a student and not a junior."

These forms can be generalized using variables for the modes of action and so we can get the following set of equivalences:

A—All S are P.	=	$\forall x\,(fx \supset gx)$
E—No S are P.	=	$\forall x\,(fx \supset \neg gx)$
I—Some S are P.	=	$\exists x\,(fx \wedge gx)$
O—Some S are not P.	=	$\exists x\,(fx \wedge \neg gx)$

Now suppose that we want to prove that syllogisms in the first figure, mood AAA are valid. We can begin by representing the syllogism as a deductive argument.

> All M are P.
>
> All S are M.
>
> So, all S are P.

1. $\forall x\,(Mx \supset Px)$ Premise

2. $\forall x\,(Sx \supset Mx)$ Premise

∴ $\forall x\,(Sx \supset Px)$

In order to prove the conclusion, we begin by picking individuals that can "test" the assertions. So we can use the postulate of Universal Instantiation to introduce a "random" individual.

$$3. \text{Ma} \supset \text{Pa} \quad 1, \text{UI}$$

$$4. \text{Sa} \supset \text{Ma} \quad 2, \text{UI}$$

Since the conclusion will be in the form of a material implication, we can use a conditional argument. First, we assume whatever we like, in this case, the instantiated version of the subject term of the conclusion.

$$5. \text{Sa} \qquad \text{Assume}$$

$$6. \text{Ma} \qquad 4, 5, \text{MP}$$

$$7. \text{Pa} \qquad 3, 6, \text{MP}$$

$$8. \text{Sa} \supset \text{Pa} \quad 5\text{-}7 \text{ CA}$$

The conclusion of the conditional argument involves two propositions that regard a particular individual. Since the individual was introduced originally by rule UI, it is assumed to be any random individual from the universe of discourse. Using rule UG, we can then generalize the conclusion of the conditional argument.

$$9. \forall x (\text{Sx} \supset \text{Px}) \quad 8, \text{UG}$$

This form can also be proven in fewer lines by using the transitivity argument form:

$$5. \text{Sa} \supset \text{Pa} \qquad 3, 4 \text{ TR}$$

$$6. \forall x (\text{Sx} \supset \text{Px}) \quad 5, \text{UG}$$

And so we can prove that the first figure, mood AAA is a valid form using the equivalence rules and postulates of the system.

This same approach can be used to prove the direct inferences discussed in 4.5. For example, the converse of "Some citizens are voters" is "Some voters are citizens." To prove that the converse follows validly from the original proposition, translate both propositions into quantified assertions of the sort we have discussed and use the original as a premise to prove the converse. Let 'C' mean 'is a citizen' and 'V' mean 'is a voter'.

$$1. \exists x (\text{Cx} \wedge \text{Vx}) \quad \text{Premise} \therefore \exists x (\text{Vx} \wedge \text{Cx})$$

$$2. \text{Ca} \wedge \text{Va} \qquad 1, \text{EI}$$

$$3. \text{Va} \wedge \text{Ca} \qquad 2, \text{C}$$

$$4. \exists x (\text{Vx} \wedge \text{Cx}) \quad 3, \text{EG}$$

The contrapositive of an A assertion can also be proven by this method. Consider "All citizens are voters" and its contrapositive "All non-voters are non-citizens."

1. $\forall x(Cx \supset Vx)$ Premise $\therefore \forall x(\neg Vx \supset \neg Cx)$

2. $Ca \supset Va$ 1, UI

3. $\neg Ca \wedge Va$ 2, DI

4. $Va \vee \neg Ca$ 3, C

5. $\neg\neg Va \vee \neg Ca$ 4, DN

6. $\neg Va \supset \neg Ca$ 5, DI

7. $\forall x(\neg Vx \supset \neg Cx)$ 6, UG

Notice, however, that the direct inferences by limitation cannot be proven by this method since the system of deduction we are using supposes the modern interpretation of the square of opposition in which categories are not guaranteed to have members. However, consider the converse (by limitation) of an A assertion, for example, "All citizens are voters." If we suppose the existential interpretation, then that there is at least one voter and at least one citizen. If this is the case, then we may add a premise (in addition to the A assertions): $\exists xCx \wedge \exists xVx$.

1. $\forall x(Cx \supset Vx)$ Premise $\therefore \exists x(Cx \wedge Vx)$

2. $\exists xCx \wedge \exists xVx$ Added to represent the existential interpretation

3. $\exists xCx$ 2, Si

4. Ca 3, EI

5. $Ca \supset Va$ 1, UI

6. Va 5, 4, MP

7. $Ca \wedge Va$ 4, 6, Con

8. $\exists x(Cx \wedge Vx)$ 7, EG

Using quantified logic, one can prove each of the valid argument forms for both interpretations of the square of opposition.

7.8   GRAPHICAL PROOFS OF VALIDITY

The valid forms of the syllogism can also be proven using the graphical method discussed in Chapter Six. In addition to the rules for the logical operators, we add four more rules to handle quantification. The first two rules parallel the quantificational equivalence rules for deduction.

$$\neg \exists xAx$$
$$|$$
$$\forall x \neg Ax$$

$$\neg \forall xAx$$
$$|$$
$$\exists x \neg Ax$$

The second pair of tree rules are parallel to the rules for universal and existential instantiations. Notice that there are no rules parallel to universal and existential generalization because the method of trees is a process of reduction to incompatible alternatives. While one might find incompatible generalizations, what makes them incompatible is that when they are instantiated, they are alternative actions that cannot both be taken by individuals.

$$\forall xAx \qquad\qquad \exists xAx$$
$$| \qquad\qquad\qquad |$$
$$Aa \qquad\qquad\qquad Ac$$

In the rule for existential quantification, 'c' stands for a new individual term (one that does not already appear in the branch). In the rule for universal instantiation, 'a' stands for any individual term that has already been introduced. It is important to remember that *a new individual term can only be introduced by instantiating an existentially quantified assertion* (e.g., ∃xAx). Universally quantified assertions can only be instantiated using individual terms that already appear in the tree. As in the case of deductive proofs, *the existential instantiation rule should be used first whenever possible.*

In trees for propositional logic, each line is checked off once the relevant rule has been applied. This is true in trees for quantified logic as well, with one significant exception. Since universal claims apply to every individual term in the universe of discourse, the universal instantiation rule can be used for every new term introduced in the tree. As in the case of propositional trees, a branch closes when propositions of the form p and ¬p both appear on the branch. The argument is proven when all of the branches close. When one or more branches remain open, there are assignments of truth-value that will make the premises true and the conclusion false and so the form is not valid.

Consider again the proof for a syllogism in the first figure, mood AAA. The tree begins in the same way as the trees for propositional proofs, by writing the premises and the negation of the conclusion.

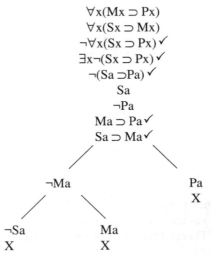

A syllogism of the third figure, EIO, can be proved as follows:

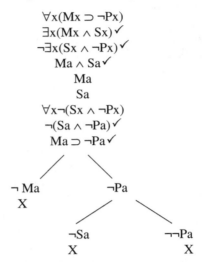

A proof of the existentially valid syllogism, fourth figure, AAI, requires the addition of a premise that asserts the existence of at least one member of each category (the italicized premises).

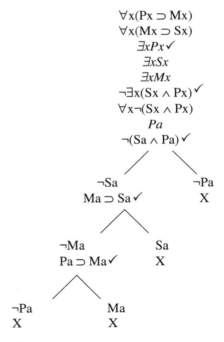

For invalid arguments (arguments with a branch that does not close), an assignment of truth values can be given in which the premises are true and the conclusion false. This is done by reading the assignment of values for each predicate as given on one open branch of the tree. Consider the following invalid argument:

$$\forall x(Ax \supset Bx)$$
$$\forall x(Ax \supset \neg Cx)$$
$$\neg\forall x(Cx \supset Bx)\checkmark$$
$$\exists x\neg(Cx \supset Bx)\checkmark$$
$$\neg(Ca \supset Ba)\checkmark$$
$$Ca$$
$$\neg Ba$$
$$Aa \supset Ba$$

```
                 Aa ⊃ Ba
                 /      \
            ¬Aa          Ba
        Aa ⊃ ¬Ca          X
         /    \
      ¬Aa      ¬Ca
                X
```

The left-most branch remains open. In this case, the premises are true and the conclusion of the argument false when, given the individual called "a," Aa is false (i.e., ¬Aa is true), Ba is false (¬Ba is true), and Ca is true.

7.9 EXERCISES

7.9.1. Prove the following syllogisms valid first using natural deduction and then using the method of tableaux:

First Figure, Moods EAE, EIO

Second Figure, Moods AEE, AOO

Third Figure, Moods AII, OAO

Fourth Figure, Moods AEE, IAI

7.9.2. Construct formal proofs for all the arguments below. Use equivalence rules, truth functional arguments, and the rules of instantiation and generalization. These may also be proven using the method of tableaux.

1. $\forall x(Ax \supset \neg Bx) \therefore \neg\exists x(Ax \wedge Bx)$

2. $\forall x(Mx \supset \neg Px), \exists x(Sx \wedge Mx) \therefore \exists x(Sx \wedge \neg Px)$

3. $\exists xAx \vee \exists x(Bx \wedge Cx), \forall x(Ax \supset Cx) \therefore \exists xCx$

4. $\forall xLx \therefore \forall x(Px \supset Lx)$

5. $\forall x(\neg Fx \supset Gx), \neg\exists xFx \therefore \forall xGx$

6. $\forall x(Cx \supset \neg Sx), Sa \wedge Sb \therefore \neg(\neg Ca \supset Cb)$

7. $\exists xCx \supset \exists x(Dx \wedge Ex), \exists x(Ex \vee Fx) \supset \forall xCx \therefore \forall x(Cx \supset Gx)$

8. $\forall x(Fx \supset Gx), \forall x[(Fx \wedge Gx) \supset Hx] \therefore \forall x(Fx \supset Hx)$

9. $\exists xLx \supset \forall x(Mx \supset Nx), \exists xPx \supset \forall x\neg Nx \therefore \forall x[(Lx \wedge Px) \supset \neg Mx]$

10. $\forall x(Fx \equiv Gx)$, $\forall x[(Fx \supset (Gx \supset Hx)]$, $\exists xFx \lor \exists xGx \therefore \exists xHx$

11. $\exists x(Cx \lor Dx)$, $\exists xCx \supset \forall x(Ex \supset Dx)$, $\exists xEx \therefore \exists xDx$

12. $\forall x[(\neg Dx \supset Rx) \land \neg(Dx \land Rx)]$, $\forall x[Dx \supset (\neg Lx \supset Cx)]$, $\forall x(Cx \supset Rx)$ $\therefore \forall x(Dx \supset Lx)$

7.9.3. Using the method of tableaux, give an assignment of values for the predicates of each argument that shows that each argument is invalid.

1. $\forall x(Ax \supset Bx)$, $\forall x(Ax \supset Cx) \therefore \forall x(Bx \supset Cx)$

2. $\exists x(Ax \land Bx)$, $\forall x(Cx \supset Ax) \therefore \exists x(Cx \land Bx)$

3. $\forall x[(Cx \lor Dx) \supset Ex]$, $\forall x[(Ex \land Fx) \supset Gx] \therefore \forall x(Cx \supset Gx)$

4. $\exists xMx$, $\exists xNx \therefore \exists x(Mx \land Nx)$

5. $\forall x[Dx \lor (Ex \lor Fx)] \therefore \forall xDx \lor (\forall xEx \lor \forall xFx)$

6. $\exists x(Cx \land \neg Dx)$, $\exists x(Dx \land \neg Cx) \therefore \forall x(Cx \lor Dx)$

7.10 BORDER AGENTS AND THE PROBLEMS OF LOGIC

In the context of his work on systems of order, Royce concluded that whenever two things (classes or actions) are distinct from each other, there emerges a third mode of action that, like Peirce's boundary, is compatible with both of the contrasting terms and remains distinct from them. The result is a generative ordering system that gives rise to terms that can be treated as self-identical and members of classes, but are nevertheless dependent on the activities of division and contrast.

Karen Barad, in her book, *Meeting the Universe Halfway* (2007), provides a contemporary theory of agency, "agential realism," that echoes the earlier theories of Royce and Peirce. According to Barad, individuals of every kind must necessarily be understood as the results of activities that connect and divide.[2] Using the quantum theory of Neils Bohr, Barad proposes that matter, the full range of things in experience, are the products of complex "intra-actions." In the case of Peirce's chalkboard (discussed in Chapter One), for example, the emergence of the black and white fields is the result of the intra-activity of making the chalk slash on the board. In this sense, the cut made on the board intra-acts with both the chalk used to make the cut and the chalkboard to generate two definable fields. The fields themselves are not static entities but rather ongoing activities that intra-act with other "things" or activities present to them. The nature of the cut or the division is not simply a formal operator indicating a division, but rather an "agential cut" (2007, 140), that is, a division brought about through a purposive activity. All three elements in the relation, the contrasting elements, x and ¬x, and the cut itself,

---

[2] Agency and purpose for Royce and others who posit purpose as a starting point for ontology also recognize consciousness as that which carries out the operations of "noting, observing, and recording" the modes of action necessary to achieve a purpose. Consciousness, in this context, is something that develops in the face of situations where purposes are yet to be achieved and gains its character in the process of transforming the situation.

are modes of activity that intersect in the dividing and defining moment. From this perspective, the identity of the term that emerges in the cutting can be understood as a boundary function, and the boundary, following Royce and Peirce, has a precisely specifiable but contradictory character. The boundary activity is at once compatible with each of the "sides" it divides even as it remains distinct from both.

When this conception of agency is taken up as a framework for understanding the experiences of Du Bois and Anzaldúa (also discussed in Chapter One), they become central. In Anzaldúa's example, the *mestiza* consciousness is an identity that emerges at once with the dualisms that frame her and, as she suggests, affords a particular kind of agency. To be on a border in Anzaldúa's sense is a particular mode of action that emerges in the presence of contradiction. The demands of Mexican culture, indigenous culture and gringo culture call for incompatible ways of acting. A *mestiza* emerges in the face of these demands at the intersection of culture, land, body, and desire. *Mestiza* identity is thus a border agent that is at once compatible with any of the intersecting modes of action (it is this compatibility that sets up the impossibility of action without choice) even as she is distinct (as the contradictory intersection) and therefore torn. The position is untenable in the sense that, as a mode of action, she cannot sit still in hopes that the world will simplify its relations. To simply wait is a torment, as she suggests, and to act—that is, to choose—is torment also tempered by the hope of change. It is this state of *mestiza* existence that provides her with the possibility of transformation both for herself and the networks of relations in which she lives. An agent ontology of the sort proposed by Royce and Barad provides a way to understand both the character of *mestiza* consciousness and its potential for transformation.[3]

Rather than excluding the possibility of agency at the border, the idea that agency is a characteristic of boundaries provides a way to account both for the experience of those who live contradictions and of those whose purposes do not lead to contradictions. For those like Du Bois and Anzaldúa, the experience of contradiction marks the process of agency in their lives. Rather than rejecting such a conception of agency on grounds of its contradictory nature, the *mestiza* and other border agents can be recognized as distinct and essential to change. Starting from this notion of agency, we can reexamine the value and methods of the intersections among communities, classes, and traditions and rethink the received view of logic that enforces a strict and hierarchical separation between classes. This notion of border agency also provides a

---

C. I. Lewis, in his logic of strict implication, identifies the presence of contradictions as affording the possibility of "explosion" or the ability to conclude anything at all from a contradiction. Priest and others argue that "explosion" is an unacceptable consequence of strict implication as a theory of the conditional relation. (See Priest, 2008, 74–7) "Explosion," they say, is counterintuitive and so counts against the theory. One might argue in reply that from the perspective of a border agent, in fact, "explosion" captures the character of a situation in which one faces incompatible imperatives. Given that no action can follow mechanically, whatever action occurs will have the character of an abduction—a guess, a creative moment—that has the potential to bring change or a new direction by "exploding" the impasse for better or worse.

means to account for the activity of so-called race traitors, straight allies of gays and lesbians, white advocates of indigenous people, and others. Such allies can be understood as themselves facing incompatible alternatives that can make new directions possible. Agency that develops at the intersection of interests, commitments, claims about the world, and so on, are part of the instability of the boundary that opens the possibility of reconstructing the horizon and the lifeworld it bounds.

This study began by proposing four problems that might benefit from the study of logic. The first, *abstraction*, raised the problem that logic helped to reinforce the development of abstract categories that were not subject to criticism in the context of experience but which provided a framework for organizing experience. As Andrea Nye argued, these abstractions were a product of a particular society that served to reinforce the privilege and power of some groups over others. Once established as abstractions and so removed from their origins and treated as part of a realm of unchanging truth, the categories and principles of logic became instruments of subjugation. Nye's own work argued that the study of logic could provide a means of undermining these instruments of oppression. In light of the study here, it might be possible to reassess Nye's challenge and propose other critical and constructive resources for thinking about abstraction.

In response to Nye's concerns, it is possible first to follow Nye in giving a genealogy of abstractions that have come to be accepted as fixed and unchanging truth. Just as Dewey in *The Quest for Certainty* provided an account of the ways in which philosophy as a mode of inquiry became disconnected from its founding problems, logic as a theory of inquiry provides a way to recognize that what came to be taken as abstract truths are products of responses to concrete problems found in particular cultural circumstances. The notions of strategic and communicative action also provide a means of critique. Abstractions, when taken as fixed truths, are useful in strategic action as a way to set aside questions and the possibility of revision in order to get others to accept things as they are. Communicative action, in contrast, provides a means of questioning abstractions, their usefulness, and their consequences and of revising them to respond to new circumstances and opportunities.

The second problem for logic was *dualism*. In part, this problem became apparent in response to Nye's work and her diagnosis of the problem of abstraction. For Nye, abstraction was a consequence of the use of negation as a means of establishing categories of meaning. Val Plumwood, in response, acknowledged that logic, as it came to be understood in Western philosophy, provided support for the subjugation of women and other categories (nature, matter, bodies, emotion, and so on), not by the use of negation alone (dichotomy), but rather by combining the acts of division necessary for meaning with presupposed hierarchical relations. In the dualism of man and woman or culture and nature, the division between the categories provides a means of identifying aspects of experience. Dualism becomes an issue when the act of division also reinforces hierarchies of value such that women are to be understood as servants of men and nature is viewed as subject to the control of culture.

In the end, Plumwood affirms the centrality and character of acts of judgment, but argues that the products of judgment are not unchanging hierarchies but divisions constrained by purpose and the horizon of a lifeworld. It is also clear that while some categories are the result of negation, other categories emerge as aspects of consistent modes of action that follow from particular judgments. Categories of gender, as Sharp (1995) points out (in Chapter One), do not need to be defined dualistically, but may be defined as a series of related categories with overlapping features that mark both division and connection. The resources of logic as we have them can provide a means for critically reconstructing the process of category formation in a way that may undermine subjugation and foster freedom for those within a given lifeworld.

The third problem, *incommensurability*, marked the difficulty human beings face in trying to understand each other and coordinate actions across significant differences in language, culture, and purpose. For some, incommensurability marks a real limit to the possibilities of human interaction and the ability of diverse human communities to find common ground. Others argue that incommensurability is only a temporary problem that can be overcome by study and interaction and that taking the problem too seriously masks the potential for universal conceptions of the good, the true and so on. While often not considered a logical problem, incommensurability turns on whether or not diverse systems of order share any commonalities in terms of which they can interact in ways that preserve differences and yet make common action and purpose possible. The difficulty, in part, is answered by the formulation of the problem by the various agents involved. Insofar as diverse systems of order are made, the principles necessary to the ordering of the issue serve as a common ground. Whether these commonalities of order provide a context for the consideration of different and even contradictory judgments and purposes remains a question for discussion.

The fourth problem, the problem of *boundaries*, provides a key for responding to the problem of incommensurability. Boundaries have appeared in many forms in the course of this study: first, they marked the distinctions between human groups in the experience of Gloria Anzaldúa and W. E. B. Du Bois. Later boundaries emerged as the horizons of situations that constrain inquiry in certain ways and define lifeworlds. Still later they became the lines between the categories necessary for syllogisms to work. When these sharp boundaries broke down in the face of the processes of category formation, they still persisted in the context of induction as experienced boundaries of relevance in light of present inquiries and purposes.

In the discussion of deductive systems, boundaries reemerged in the context of acts of judgment. Rather than being passive markers of difference, boundaries are modes of action that stand between incompatible alternatives and as such can be recognized as the fundamental logical operation: exclusive disjunction. From this perspective, the emergence of different courses of action turned on the activities of agents in the face of forced options. On closer examination, the act of judgment proved to have a structure that could be formalized in a way that can provide a notion of formal validity and a means of reframing courses of action, trying alternatives, and criticizing the results. Such formal systems, however, remain formal and carry with them the risk of abstraction. At the same time, they provide a way to study the character of the interactions that emerge in the context of forced options and provide insights that can be used in thinking again about the processes of category formation, valuation, critical inquiry, and social change.

In fact, it would seem that an understanding of boundaries helps to answer each of the other problems of logic. For example, boundary agents can overcome incommensurability by their very nature and can unify seemingly distinct qualities in a way that undermines particular dualisms. Boundaries even provide a means for rethinking abstraction. While abstractions are separated from particulars, often for the purpose of making sense of them, what divides an abstraction from the particular that suggests it is the interest of the agent trying to make sense of the experience at hand. From this perspective, abstractions never really were abstract in the traditional sense but are the products of border agents who stand between the present situation and its future.

It remains an ongoing project to think about logic, argument, and the science of order and the ways in which it can (and cannot) foster a better understanding of judgment, ethics, social relations, and knowledge and open further areas of investigation that can address new problems in the 21st century.

Solutions

CHAPTER ONE: THE SIGNIFICANCE OF LOGIC

1.5.

1. This is an example of the problem of *abstraction*. Russell advocates removing what he sees as subjective content (e.g., emotional commitment to one's country) from what he takes to be the objective form of the problem in order to arrive at the best solution. The danger here is that this might also cause us to abstract certain elements that are essential to correctly dealing with the problem (e.g., race, gender, class, age).

3. This is an example of the problem of *dualism*. For Aquinas, there is both a division and a hierarchy between positive and negative enunciations.

6. Cajete argues that Native culture and modern society are working from fundamentally different worldviews (i.e., paradigms) so that each one's interpretation of democracy may be incomprehensible to the other. This is an example of the problem of *incommensurability*.

8. This is an example of the problem of *abstraction*. Dussel argues that Kant's criterion that maxims that govern action can be universalized rests on the idea that they can be divorced from any particular experience. This is a problem for Dussel because it can permit the elevation of one particular view of the "Good" by presenting itself as universal.

13. This is an example of the problem of *boundaries*. According to Zack, the commonly held view is that racial identity is an important element of one's own identity and this is typically passed down through one's family. A problem, or question, arises for "mixed-race" identities in the United States where one is either white or non-white and biracial families are not typically accepted. It is also not clear how first-generation mixed-race people could have an identity at all.

1.7.

2. **Conclusion:** A universal is not a substance.
   Premise: Every substance is either one and not many or many.
   Premise: A universal is not singular (i.e., not "one and not many").
   Premise: A universal is not many.

   The first premise is supported by a brief argument: A universal applies to everything or a whole class of things. If Socrates is a universal, then so are

Parmenides and Protagoras. But there is no reason to choose one or the other of these as the universal (the character which applies to the whole class), so Socrates is no universal. Since Socrates was selected at random (and the same can be said of any individual selected), then we must conclude that a universal is not singular.

The second premise is supported by another argument: If a universal is many, then it is either many singular things or many universal things. It cannot be many singular things, since singular things are not universal (as we have seen). So, the universal must be many universals. But if this is so, then every universal is made up of universals, and so on *ad infinitum*. So a universal cannot be many either.

3. **Conclusion:** We [Seneca] believe it wrong for you [white missionaries] to attempt further to promote your religion among us, or to introduce your arts, manners, habits and feelings . . . [and] for us to encourage you in doing so.
Premise 1: The introduction of Christianity has so far rendered us uncomfortable and miserable.
Premise 2: The Indian youth taken and educated in white schools became discouraged and dissipated—despised by the Indians, neglected by the whites, and without value to either.
Premise 3: The failure of these first attempts at educating Indian youth was attributable to miscalculation, and we were invited to try again . . . [but] the result has been invariably the same.
Premise 4: If all this is true, then it must also follow that the Great Spirit made the whites and the Indians . . . for different purposes.

In this example, Sagoyewatha offers a series of reasons of different sorts to support the Seneca conclusion. The last premise draws directly on the missionaries' conception of God.

5. **Conclusion:** Different treatment of women and men explains the apparent difference in the cognitive abilities of women and men in general.
Premise 1: An arbitrarily divided group that is given different cognitive exercises will develop different cognitive abilities.
Premise 2: (unstated) Women and men have been divided up arbitrarily.
Premise 3: Men have been given more opportunities to exercise their cognitive abilities and brain functions.

7. **Conclusion:** The white man and the Indian see different realities.
Premise 1: The white man would say [with regard to the prairie] "There is nothing here."
Premise 2: The Indian would say [with regard to the prairie] "Everything is here."
Premise 3: . . . there is no agreement [between the white man and the Indian] on what is "here" even though both see the same phenomena.
Premise 4: (unstated) What is "here" is reality.

9. **Conclusion:** Native American spirituality is being reduced to a commodity, seized and sold.

Premises: The premises are a series of examples of elements of Native American culture seized and sold.

11. **Conclusion:** There is a problem for revolutionary struggle in defining power in terms of binary structures: possessing power versus being powerless.
Premise 1: Relative to the status of women, they are understood as powerless unified groups.
Premise 2: If the struggle for a just society is a movement from powerlessness to power, then the resulting society simply reverses the relation: those who were powerful become the powerless.
Premise 3: As a model of bringing about a just society, this assumes that women will use power differently.
Premise 4: But women are not essentially superior or infallible.

The problem is that if we understand the struggle for justice in terms of power alone and women are not essentially superior or infallible, the result of the struggle may not be a more just society.

13. **Conclusion:** History is constantly being rewritten.
Premise 1: Conceptual material used to write history is found at the time the history is written.
Premise 2: As cultures change, these conceptual materials must also change.
Premise 3: New conceptual material leads to new data included.

15. **Conclusion:** The crises of Western culture (a loss of meaning) can be addressed by generating new ideas that "coordinate with reality."
Premise 1: When ideas are not coordinated, they lose their meaning (or, when they have meaning, then they are coordinated with reality).
Premise 2: Old ideas become artifacts and make no sense to the common people.
Premise 3: So new ideas are needed that will "justify reality once again."

17. **Conclusion:** Locke was mistaken in rejecting all innate illuminations.
Premise 1: Locke admits that some ideas come from reflection.
Premise 2: Reflection is attention to what is within (i.e., what is innate).
Premise 3: The senses do not give us what we carry within.
Premise 4: There are many ideas that are innate.

19. **Conclusion:** White Southern culture in the 1950s and 1960s faced a dilemma: outsiders must be kept out of Southern society and outsiders must be admitted to Southern society.
Premise 1: Everything was fine before the outsiders (anti-segregation activists) interfered.
Premise 2: To admit outsiders would create an intolerable situation and so they must be excluded.
Premise 3: To admit that the strife was the result of insiders (Blacks in the South) was impossible.
Premise 4: So, to explain the unrest, the unruly outsiders had to be admitted.

CHAPTER TWO: WHAT IS LOGIC?

2.4.2.

1. I began by reading the problem and saw that I needed to figure out what color hat the Green Party candidate wore (given that she won). I decided that the only way one of the candidates would answer is if the candidate were certain of the color of her or his hat. If the candidate saw two black hats, then she would certainly know that hers was white. If she saw one black hat and one white or two white hats, her hat could be either black or white. Since the Green candidate answered, her hat had to be white.

3. Setting aside the context, I focused on how the two priests might answer the question. If the leader asks the liar which door would lead him home, then the lying priest would have to lie and name the door that leads to the bear. If the leader asks the honest priest, the priest will tell the truth. Unfortunately, the leader doesn't know which priest is which. In order to get the priests to answer the same way, the leader could ask either priest, "If I ask the other priest which door will lead to my freedom, which door would he pick?" The lying priest would have to lie and say that the honest priest would name the door with the bear. The honest priest would tell the truth and say that the lying priest would also name the door with the bear. So the leader can ask either priest the question and then choose the other door. Of course, in context, asking which door will lead to freedom will depend upon the priests' theology since death (on the view of some) is freedom.

5. I started by guessing that Barry might be a Cretan. If he is, then it is not the case that Abigail said that she is a Puritan (since Barry must be lying). If Barry lied, then Abigail must have said that she is a Cretan. But if she did, then she would be telling the truth and so could not be a Cretan. So Abigail is a Puritan. Since Barry said that Abigail said that she is a Puritan and she is (according to the hypothesis), then Barry is telling the truth and must be a Puritan too. Carl, however, declares that Abigail is not a Puritan and he knows it, which is a lie. So Carl is a Cretan.

7. The key here is to come up with the grid used to code the original message. Assuming all the cells of the grid were filled, I began by counting the letters in the code (33). In order to divide 33 with no remainder (so all the cells will be full), the options are 11 or 3. I first wrote the letters out in 11 rows of 3, but the result was a seemingly random series of letters. So I wrote the letters out in the given sequence in three rows of 11. I then read the letters top to bottom, beginning with the top left letter. The result was a quote from John Dewey:

THINGSAREWHATTHEYAREEXPERIENCEDAS

CHAPTER THREE: COMMUNICATIVE ACTION

3.2.1.

Dialogue 1:

EMIL When you vote, you should vote for me.

SUSAN I would like to vote for the best candidate.

EMIL   That's me. If you vote for me, I will see that you get free tickets to next week's football game.

SUSAN   How does that make you the best candidate for the student government executive board?

EMIL   Because you benefit from my election.

Dialogue 2:

ROBERTA   If I'm elected to the student government executive board, what stand should I take on raising student fees?

ARTHUR   Fees have gone up a lot in the past couple of years. Maybe you should favor a freeze on further increases for one or two years.

ROBERTA   That may make it hard to adequately fund student organizations.

ARTHUR   True, but if they are increased further it will make it hard for students to afford books and even tuition.

ROBERTA   I've heard the same concern from several other students. I think I'll support a freeze.

In the first dialogue, Emil the candidate wants Susan to vote for him. Rather than giving reasons that address Emil's candidacy on its merits, and so to seek Susan's agreement as a result of the conversation, he offers her free tickets. When she questions the free tickets as a reason, he tries to make Susan see the advantage to her of his election (again, not in virtue of the conversation, but in virtue of the tickets). Emil's speech acts are strictly aimed at realizing a result independent of the conversation at hand. In contrast, the second dialogue seeks to find some agreement about what Roberta should do in office. The dialogue marks an exchange of information in which both speakers participate. Agreement is arrived at as a result of the conversation not—as in the first dialogue—because of something independent of the conversation (i.e., the free tickets).

3.2.2.

1. Since communicative action necessarily involves interaction in which the participants can ask questions and change their minds, it seems that no printed speech could be an instance of communicative action. On the other hand, a printed speech can be part of a larger exchange in which the speech serves the function of proposing some of the ideas that provide a starting point for further exchanges. Such speeches can even overtly set the stage for questioning and the possibility of non-coerced belief. In this section of her speech Addams, in fact, challenges the idea that beliefs can be coerced and takes as her example bayonet charges across "no-man's land." Addams' claim is that apparently "heroic" charges were not a matter of belief or loyalty but a result of the systematic use of drugs on the soldiers of all sides. Rather than demanding a conclusion in light of the evidence, she actually invites the audience to consider the implications of the information: "Well, now, think of that." The next paragraph raises the issue of responsibility for the war. She invites the audience to consider the issue in the context of "human nature" reasserting itself in the form of a question: why didn't you (her American audience) take action against "this horrible thing"?

Until her speech at Carnegie Hall, Addams had been regarded as one of America's leading activists. Following her speech, she became the target of attacks in the press, her social work projects were threatened, and she was even brought before a congressional committee as a possible traitor for her peace work in Europe. Most critics cited these passages as the reason for their condemnation. It seems that her claim that soldiers needed to be drugged in order to make them charge into near certain death showed that she failed to support the troops. After the war, her claims about the use of drugs were confirmed and she was awarded the Nobel Peace prize.

5. The dialogue between Philonous and Hylas, from one angle, seems to be a clear case of strategic action. Philonous, here and in the dialogue as a whole, rarely seems to be asking Hylas a real question and rarely seems to listen to the answers he receives. To the extent that Berkeley's dialogue is presenting a record of the dialogue, the dialogue is better characterized as a strategic action on the part of Philonous to force Hylas to agree with a position that he finds absurd. At the same time, the dialogue is not simply a record of a conversation but a contribution to a larger conversation at the time (and since) on the nature of experience and reality. In this case, Philonous's strategic questioning is also the presentation of a particular view that invites objection from the reader. The dialogue form recalls the philosophical method of Socrates and, for an 18th century reader, the method of learning used by Christian catechisms. This first kind of dialogue marks (or was taken to mark) a process of open inquiry in which the participants formed their beliefs without coercion. The second kind of dialogue used the method of questions and answers as a means of clearly stating doctrine as a precursor to belief (and, potentially, to debate). The dialogue is an instance of strategic action from one angle and communicative action from another.

3.4.2.

1. Addams on War

Foreground Knowledge

Situation-Specific Horizontal Knowledge: In a sense, Addams and her colleagues, delegates from the International Conference of Women for a Permanent Peace held at the Hague, were involved in a fact-finding mission in order to assess the best way to work for peace. Most of the audience attended the speech in order to hear the results of her trip. In this case, most of the audience was already familiar with particulars of the conflict, the nations involved, the leaders of those nations, and so on. Most of the audience would have known that the United States was officially neutral in the conflict, though there was a growing number of people arguing that the U.S. should enter the war even as there was a growing peace movement. The horizon of the speech was formed by the various debates about the role of the U.S. and the morality of intervention and of war in general.

Topic-Dependent Contextual Knowledge: The reference to the use of drugs to coerce behavior and to "human nature" suggests that audience brought a range of assumptions about the issues at stake. The implication of Addams' claims relied on the expectation that their knowledge would allow them to make some judgment about the issues at hand.

Background Knowledge

In the background of Addams' speech there is a wide range of expectations about what is permissible and what is not permissible in the face of war. The response to Addams' speech demonstrated that there were ideas and expectations in the background that were unquestioned and came to frame the understanding of her claims in a way that prevented ongoing communicative action. At the same time, Addams clearly expected that her audience would recognize the evil of war and would take seriously and thoughtfully any honest attempt to end it.

5. Berkeley's Dialogue

Foreground Knowledge

Situation-Specific Horizontal Knowledge: In the context of philosophical writing, the horizon of the discussion is probably best understood as formed by the philosophical questions of the time. In this case, the problem of understanding the senses is presumed. Hylas and the readers of the time do not need to be convinced that it is important to develop a doctrine of the senses that accounts for cases such as the experiment in which something feels both hot and cold.

Topic-Dependent Contextual Knowledge: The question at hand and the discussion of the temperature experiment are connected with a wider set of principles and debates of the time that both speakers (and the readers) would know. At stake in the excerpt is a set of principles about the character of matter. A response to the problem recognized by the speakers would be either a way of accounting for the experiment in terms of the received theory of matter or in finding a better theory.

Background Knowledge

In this case, the background knowledge includes the expectation that philosophical debate matters and that it is be carried out in a way that recognizes certain things as true and false and that wider logical principles (the idea that principles that lead to contradictions are absurd). It also includes (especially in the context of the full debate between Philonous and Hylas) that God exists and has a certain character that must also be taken into account in any theory of matter and perception.

3.7.

1. A fallacy of postulates, *petitio principii.* The mayor is citing her own judgment as the reason for accepting her judgment.

3. A fallacy of the problem, false dilemma. The problem is presented as a choice between two things which do not exhaust the possibilities.

5. A fallacy of postulates, composition. The speaker attributes recovery to the entire salmon run because one fish survived.

7. Fallacy of postulates, composition. Mill offers something true of a part as evidence for a truth about the whole (that a part could be happy means that there must be a comparable happiness for the whole).

9. Fallacy of connection, *ad populum*. In this case, the speaker takes popularity (or unpopularity) to support a claim about what the policy might realize—in effect, popularity is irrelevant to the claim to be supported.

11. Fallacy of connection, solipsism. The speaker disregards opposed interpretations of the evidence in favor of her interpretation.

13. Fallacy of postulates, source (prejudice). The owner selects the evidence to be considered based on a prejudice about the worker. The owner ignores the possibility of the customers as the culprits.

15. Fallacy of connection, *ad misercordium*. Ella draws her conclusion on the basis of her sympathy for the developer. One might also call it an *ad hominem* fallacy, though, in this case, rather than rejecting a conclusion based on a person's character, we are asked to accept it.

17. Fallacies of connection, *ad misercordium* and *ad baculum*. The press secretary first appeals to pity (he's very tired) and then to a threat (if you don't agree, bad things will happen).

19. Fallacy of the premises, equivocation. The speaker speaks as though the only moral issue involves the shareholders and not the viewers.

21. This is a fallacy of *ambiguity* (composition). Even though all the parts are great, that is, great writer, actors, director and producers, it does not mean that the whole will be great.

23. This is a case of a fallacy of connection, *ad misericordium*. While there might be some equivocating going on with the term "deserve" the real force behind this argument is the sympathy aroused by Sean's present predicament.

25. This is a case of a fallacy of connection, *ad baculum*. The persuasiveness of the argument is not so much in its reasoning as it is in the fear (e.g., vandalized shops) that it invokes.

27. This is a case of a fallacy of the conclusion, *post hoc, ergo propter hoc*. This argument assumes that because pitchers stop pitching well after they have been talked to in-between innings that talking to pitchers in-between innings causes them to stop pitching well. This is not to say that the theory is not good, but the argument does not show why or how the two are causally connected, it merely states that one follows the other and therefore one causes the other.

29. This is a case of a fallacy of validity, denying the antecedent, because, while it is true that taking care of yourself when you are young might contribute to your living a long life, it is possible that you could live a long life for other reasons.

31. This is a fallacy of exclusion, complex question, because it conjoins two things which do not necessarily go together, that is, voting for a democrat and

running a state into financial ruin, and therefore answering either yes or no to this question would in fact affirm that problematic conjunction.

33. This is a fallacy of connection, solipsism, since the conclusion considers only the speaker's point of view on the matter.

35. This is a case of a fallacy of the conclusion, irrelevant conclusion. The premise of the argument may be accurate, but it is not clear how the conclusion follows.

CHAPTER FOUR: THEORY OF THE SYLLOGISM

4.4.

1. Some good ideas are things needed to help get this off the ground

3. All baseball fans that watch the World Series are baseball fans with cable.

5. All academics who care about the Good are philosophers.

7. Some professional athletes are not fast runners.

9. Some voters are not those going to vote for a major party candidate.

11. No acts of taxation are acts without representation.

13. All asbestos-contaminated buildings are serious cancer risking things.

15. The writer of this song is a person who had a big impact on everyone here.

17. Some patriotic Americans are not voters in presidential elections.

19. Some flowers are things that bloomed last week.

4.4.2.

1. **Subaltern**: Some of the players on the basketball team are over six feet tall.
   Contrary: No players on the basketball team are over six feet tall.
   Contradictory: Some of the players on the basketball team are not over six feet tall.

3. **Subaltern**: Some Martians are not things that exist.
   Contrary: All Martians are things that exist.
   Contradictory: Some Martians are things that exist.

5. **Subaltern**: All pro-athletes are out of shape.
   Contrary: Some pro-athletes are not out of shape.
   Contradictory: No pro-athletes are out of shape.

7. **Subaltern**: No famous artists are talented.
   Contrary: Some famous artists are talented.
   Contradictory: All famous artists are talented.

4.6.1.

1. **Converse**: Some things needed to get off the ground are good ideas.
   Obverse: Some good ideas are not non-things needed to help get this off the ground.
   Contraposition: None

3. **Converse:** (limitation) Some baseball fans with cable are baseball fans that watch the World Series.
Obverse: No baseball fans that watch the World Series are non-baseball fans with cable.
Contraposition: All non-baseball fans with cable are non-baseball fans that watch the World Series.

5. **Converse:** (limitation) Some philosophers are academics who care about the Good.
Obverse: No academics who care about the Good are non-philosophers.
Contraposition: All non-philosophers are non-academics who care about the Good.

7. **Converse:** None
Obverse: Some professional athletes are non-fast runners.
Contraposition: Some non-fast runners are not non-professional athletes

9. **Converse:** None.
Obverse: Some voters are those not going to vote for a major party candidate.
Contraposition: Some of those not going to vote for a major party candidate are not non-voters.

11. **Converse:** No acts without representation are acts of taxation.
Obverse: All acts of taxation are acts not without representation.
Contraposition: (limitation) Some acts not without representation are not acts of non-taxation.

13. **Converse:** (limitation) Some serious cancer risking things are asbestos-contaminated buildings.
Obverse: No asbestos-contaminated buildings are non-serious cancer risking things.
Contraposition: All non-serious cancer risking things are non-asbestos-contaminated buildings.

15. **Converse:** (limitation) Some person with a big impact on everyone here is the writer of this song.
Obverse: No one who wrote this song is a person who had a big impact on everyone here.
Contraposition: All persons who have not had a big impact on everyone here are non-writers of this song.

4.6.2.

1. 1) No smokers are healthy people (premise).

2) No healthy people are smokers (converse of 1).

3) All healthy people are non-smokers (obverse of 2).

4) All non-healthy people are non-non-smokers (contraposition of 3).

3. 1) All socialists are good comrades (premise).

 2) No socialists are non-good comrades (obverse of 1).

 3) All socialists are non-non-good comrades (obverse of 2).

 4) No socialists are non-non-non-good comrades (obverse of 3).

5. 1) Some tasty drinks are sugary beverages (premise).

 2) Some tasty drinks are not non-sugary beverages (obverse of 1).

 3) Some non-non-sugary beverages are not non-tasty drinks (contraposition of 2)

 4) Some non-non-sugary beverages are non-non-tasty drinks (obverse of 3).

7. 1) All places where one should shop are small businesses (premise).

 2) All non-small businesses are non-places where one should shop (contraposition of 1).

 3) No non-small businesses are non-non-places where one should shop (obverse of 2).

 4) No non-non-places where one should shop are non-small businesses (converse of 3).

4.8.1.

1. No Republican is an advocate of socialized medicine.

All people in favor of small government are Republicans.

So, no people in favor of small government are advocates of socialized medicine.

First Figure, Mood EAE. Valid.

2. All people who know their subjects well are good teachers.

Some people who know their subjects well are scientists.

So, some scientists are not good teachers.

Note: "Only some scientists" is read as some scientists are *and* some are not. Since this is not a standard form proposition, one approach is to give two syllogisms.

The first, however, using 'Some are not', is third figure, AIO, invalid.

The second (using 'Some are') is third figure, AII, valid.

3. All barbers I have met are excellent musicians.

Some politicians I have met are not excellent musicians.

So, some politicians I have met are not barbers.

Second figure, AOO, valid.

4.8.2.

1. First figure, Mood AAA, Valid

3. Fourth figure, Mood EAI, Invalid

5. Second figure, Mood EIO, Valid

7. Third figure, Mood IAI, Valid

9. Third figure, Mood EAE, Invalid

11. Fourth figure, Mood AEE, Valid

CHAPTER FIVE: INDUCTION AND THE LIMITS OF REASON

5.4.

1. Invalid. Fails to be an appropriate sample because, even though the sample is random and objective, it does not tell us about who opposes the war in Iraq (e.g. people could be indifferent or conflicted over the issue).

3. Valid. The conclusion might seem to be a bit of a leap, but the sample appears to be random, appropriate, and objective.

5. Valid. The comparison between groups (financial institutions and children in this case) is common. Given certain assumptions about the two groups (that they tend to be self-interested and have little self-control), the comparison seems appropriate and the conclusion follows from the analogical premises.

7. Invalid. While the study might be good for many purposes and various uses it is probably not appropriate as the basis for a conclusion about students in my logic class, if only because the financial circumstances of college students in general are better than those of the population at large.

9. It is debatable whether this well-recognized comparison is valid or invalid. Given the differences between the individual and the state relative to the notion of justice, one might conclude that the comparison is invalid. In any case, the example shows the value of analogies in raising questions and deciding on issues of relevance.

11. Invalid. While the source might be reliable and correct, this information is not appropriate to this case. The confusion seems to come from the two different uses (i.e., senses) of 'average American household'.

5.6.1.

1. Necessary. It is necessary that a U.S. citizen turn 18 years of age before she or he is eligible to vote, but being an 18-year-old U.S. citizen is not sufficient to be eligible to vote.

3. Sufficient. There are many ways to be convicted of a crime, but being an accomplice to a burglary is sufficient.

5. Sufficient. A person can become a citizen by the "naturalization" process and so it is not necessary to be born in the United States to be a citizen. If one is born in a U.S. state, however, it is sufficient to be citizen.

7. Both necessary and sufficient. According to these statutes to be Black it was both necessary and sufficient for a person to have any trace of African ancestry.

5.6.2.

1. *Method of Residue.* The example shows a case where two of the possible factors are taken into account and eliminated from consideration, but the phenomenon continues. It is therefore reasonable to conclude that it is the remaining factor that is the root cause.

3. *Method of Difference.* In this case there is agreement on most of the critical issues for re-election. The primary difference was that Stevens was convicted a week before the election.

5. *Method of Residue.* This example shows a case where all but one of the possible factors that could lead the community to have hard feelings towards the city are addressed, except for the lost sense of community. It is therefore reasonable to conclude that it is this remaining factor that is the root cause of the community's continued hard feelings towards the city.

7. *Combination of Difference and Agreement.* In this example, we see that where San Ysidro and La Jolla differ is also where La Jolla and Torrey Pines are similar—in the areas of wealthy students and the surrounding real estate value. For this reason we can attribute to these factors the cause of why students from La Jolla and Torrey Pines tend to have a higher rate of graduation and are more likely to go to a four-year university than are the students of San Ysidro.

5.8.1.

1. Classical theory and the probability is 1/4 or 0.25 (25%).

3. Relative frequency theory and the probability is 15/100 or 0.15 (15%).

5. Subjective theory and the probability is 2/3 or 0.67 (67%).

7. Subjective theory and the probability is 4/5 or 0.8 (80%).

9. Subjectivist theory and the probability is 1/10 or 0.1 (10%).

5.8.2.

1.

a) *Inclusive Disjunction Principle.* The chance of a progressive winning the race for mayor is 1/2 (as there are two candidates and the progressive is only one of them). The same, 1/2, is true of the race for city attorney. Since it is possible that a progressive can win both races we have an inclusive disjunction as follows:

$$[(1/2)+(1/2)]-[(1/2)\times(1/2)]=(2/2)-(1/4)=1-0.25=0.75\,(75\%)$$

The probability that a progressive wins at least one race is 0.75 (75%).

b) *Limited Conjunction Principle.* The chance of a conservative winning the race for mayor is 1/2 (as there are two candidates and the conservative is only one of them). The same, 1/2, is true of the race for city attorney. Since the results of the two races are independent of each other and we want to see what the chances are that a conservative wins both, we have a limited conjunction that sets up in the following way:

$$(1/2)\times(1/2)=1/4=0.25\,(25\%)$$

The probability that a conservative wins both races is 0.25 (25%).

c) *Principle of Negation.* To figure out this probability we first we need to figure out what the chances are of a progressive winning at least one race and then subtract that from 1 (remember that the negation rule works by subtracting the positive outcomes from the total possible outcomes). In order to get the probability of a progressive winning at least one race we need to use the *Inclusive Disjunction Principle*. The chance of a progressive winning the race for mayor is 1/2 (as there are two candidates and the progressive is only one of them). The same, 1/2, is true of the race for city attorney. Since it is possible that a progressive can win both races we have an inclusive disjunction which sets up in the following way:

$$[(1/2)+(1/2)]-[(1/2)\times(1/2)]=(2/2)-(1/4)=1-0.25=0.75\,(75\%)$$

From here we take the 0.75 and subtract it from 1:

$$1-0.75=0.25\,(25\%)$$

The probability that a progressive wins no seats is 0.25 (25%).

3.

a) *Principle of Negation.* If we want to know the probability that the first song played will be a *Radiohead* song we need to use the principle of negation on non-*Radiohead* songs. This is because in this case we need to subtract the negative outcomes from the total possible outcomes. The total of songs in the iPod is 100 and there are 20 songs from *Radiohead* (i.e., there are 80 non-*Radiohead* songs in the iPod). This therefore sets up in the following manner:

$$1-(80/100)=1-0.8=0.2$$

The probability that the first song played will be a Radiohead song is 0.2 (20%).

b) *Limited Conjunction Principle.* If we want to know the probability of the first two songs being from *Rage Against the Machine* with a shuffle mode that lets songs repeat, we are looking for a limited conjunction. This is because the two events are independent of each other. We know that the total of songs in the iPod is 100 and there are 40 songs from *Rage Against the Machine*. This therefore sets up in the following manner:

$$(40/100)\times(40/100)=1{,}600/10{,}000=16/100\,(0.16)$$

The probability of the first two songs being from Rage Against the Machine is 0.16 (16%).

c) *Exclusive Disjunction Principle.* If we want to know the probability that the first song played will be either from *Tool* or from *System of a Down*, what we want is an exclusive disjunction. This is because we want either event to happen, but cannot have both and may have neither. The total of songs in the iPod is 100 and there are 10 songs from *Tool* and 30 songs from *System of a Down*. This therefore sets up in the following manner:

$$(10/100)+(30/100)=40/100=4/10=0.4\,(40\%)$$

The probability of the first song played being from either Tool or System of a Down is 0.4 (40%).

d) *Inclusive Disjunction Principle.* If we want to know the probability of one of the first two songs being from *Rage Against the Machine*, we are looking for an inclusive disjunction. This is because we want either of two events to happen (e.g., a *Rage Against the Machine* song playing first or playing second), but it is just as good if both events happen. The total of songs in the iPod is 100 and there are 40 songs from *Rage Against the Machine*. This therefore sets up in the following manner:

$$[(40/100) + (40/100)] - [(40/100) \times (40/100)]$$
$$= (80/100) - (1,600/10,000) = 0.8 - 0.16 = 0.64 \, (64\%)$$

The probability that one of the first two songs will be from Rage Against the Machine is 0.64 (64%).

e) *Principle of Total Probability.* If we want to know the probability of the first two songs being from the same artist, we are looking for a total probability. This is because we need both songs to be from the same artist (i.e., a limited conjunction), but also that can be any of the five artists (exclusive disjunction). The total of songs in the iPod is 100 and there are 40 songs from *Rage Against the Machine*, 20 songs from *Radiohead*, 10 songs from *Tool*, and 30 songs from *System of a Down*. This therefore sets up in the following manner:

$$[(40/100) \times (40/100)] + [(20/100) \times (20/100)] + [(10/100) \times (10/100)] +$$
$$[(30/100) \times (30/100)] = (1,600/10,000) + (400/10,000) +$$
$$(100/10,000) + (900/10,000) = (16/100) + (4/100) + (1/100) + (9/100)$$
$$= 30/100 = 0.3 \, (30\%)$$

The probability that the first two songs will be from the same artist is 0.3 (30%).

f) *General Conjunction Principle.* If we want to know the probability of the first two songs being from *Rage Against the Machine* with a shuffle mode that does not let songs repeat, we are looking for a general conjunction. This is because the events in this case are not independent of each other. We know that the total of songs in the iPod is 100 and there are 40 songs from *Rage Against the Machine*. This therefore sets up in the following manner:

$$(40/100) \times (39/99) = 1,560/9,900 = 156/990 = 0.158 \, (15.8\%)$$

The probability of the first two songs being from Rage Against the Machine is 0.158 (15.8%).

CHAPTER SIX: PRINCIPLES OF ORDER AND DEDUCTION

6.7.1.

1.

1. D ⊃ F
2. D ∧ E   ∴ F
3. D   2, Si
4. F   1, 3 MP

3.

1. $(P \land A) \supset D$
2. A
3. $D \supset (R \land \neg N)$
4. P $\therefore \neg N$
5. $P \land A$ 2, 4 Con
6. D   1, 5 MP
7. $R \land \neg N$ 3, 6 MP
8. $\neg N$ 7, Si

5.

1. $C \mid H$
2. $S \supset H$
3. $P \supset S$
4. P $\therefore \neg C$
5. S   3, 4 MP
6. H   2, 5 MP
7. $(C \mid H) \land H$ 1, 6 Con
8. $\neg C$ 1, 6 Ex

6.7.2

1.

1. $A \mid B$ $\therefore \neg B \lor \neg A$
2. $A \mid \neg\neg B$ 1, DN
3. $A \supset \neg B$ 2, MI
4. $\neg\neg B \supset \neg A$ 3, CP
5. $B \supset \neg A$ 4, DN
6. $\neg B \lor \neg A$ 5, DI

3.

1. $(A \supset B) \mid C$
2. C $\therefore A$
3. $[(A \supset B) \mid C] \land C$ 1, 2 Con
4. $\neg(A \supset B)$ 3, Ex
5. $\neg(\neg A \lor B)$ 4, DI
6. $\neg\neg A \land \neg B$ 5, DM
7. $\neg\neg A$ 6, Si
8. A   7, DN

5.

1. $\neg(A \supset B)$ $\therefore A$
2. $\neg(\neg A \lor B)$ 1, DI
3. $\neg\neg A \land \neg B$ 2 DM

4. ¬¬A   3, Si
5. A   4, DN

7.

1. A ⊃ (B ∧ C)
2. ¬B   ∴ ¬A
3. ¬B ∨ ¬C   2, Add
4. ¬(B ∧ C)   3, DM
5. ¬A   1, 4 MT

9.

                                    ∴ (G ⊃ H) ∨ (H ⊃ G)
1. ¬[(G ⊃ H) ∨ (H ⊃ G)]   Assume
  2. ¬(G ⊃ H) ∧ ¬(H ⊃ G)   1, DM
  3. ¬(G ⊃ H)   2, Si
  4. ¬(¬G ∨ H)   3, DI
  5. ¬¬G ∧ ¬H   4, DM
  6. ¬¬G   5, Si
  7. ¬(H ⊃ G)   2, Si
  8. ¬(¬H ∨ G)   7, DI
  9. ¬¬H ∧ ¬G   8, DM
 10. ¬G   9, Si
 11. ¬¬G ∧ ¬G   6, 10 Con
12. (G ⊃ H) ∨ (H ⊃ G)   1-11, RA

11.

1. ¬(A ⊃ B) ⊃ C
2. ¬C ∧ ¬B   ∴ ¬A
3. ¬C   2, Si
4. ¬¬(A ⊃ B)   1, 3 MT
5. A ⊃ B   4, DN
6. ¬B   2, Si
7. ¬A   5, 6 MT

13.

1. A ⊃ B
2. C ⊃ D
3. (B ∨ D) ⊃ E
4. ¬E   ∴ ¬(A ∨ C)
5. ¬(B ∨ D)   3, 4 MT
6. ¬B ∧ ¬D   5 DM
7. ¬D   6 Si
8. ¬C   2, 7 MT

9. ¬B	6 Si
10. ¬A	1, 9 MT
11. ¬A ∧ ¬C	8, 10 Con
12. ¬(A ∨ C)	11 DM

15.

1. D ∨ ¬A		
2. ¬(A ∧ ¬B) ⊃ ¬C		
3. ¬D		∴ ¬C
4. ¬¬C	Assume	
5. ¬¬(A ∧ ¬B)	2, 4 MT	
6. A ∧ ¬B	5 DM	
7. A	6 Si	
8. ¬¬A	7 DN	
9. D	1, 8 DA	
10. D ∧ ¬D	3, 9 Con	
11. ¬C	4, 10 RA	

17.

1. ¬A		
2. (A ∨ B) ≡ C		
3. ¬B		∴ ¬(C ∧ D)
4. [(A ∨ B) ⊃ C] ∧ [C ⊃ (A ∨ B)]	2 B	
5. C ⊃ (A ∨ B)	4 Si	
6. ¬A ∧ ¬B	1, 2 Con	
7. ¬(A ∨ B)	6 DM	
8. ¬C	5, 7 MT	
9. ¬C ∨ ¬D	8 Add	
10. ¬(C ∧ D)	9 DM	

19.

1. A ∨ B		
2. C		
3. (A ∧ C) ⊃ D		∴ D ∨ B
4. ¬D	Assume	
5. ¬(A ∧ C)	3, 4 MT	
6. ¬A ∨ ¬C	5 DM	
7. ¬¬C	2 DN	
8. ¬A	6, 7 DA	
9. B	1, 8 DA	
10. ¬D ⊃ B	4, 9 CA	

11. ¬¬D ∨ B   10 DI

12. D ∨ B   11 DN

6.7.3.

5.

¬(A ⊃ B)✔

¬A

A

¬B

X

7.

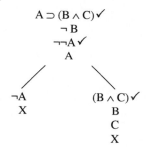

A ⊃ (B ∧ C)✔

¬ B

¬¬A✔

A

¬A   (B ∧ C)✔

X   B

C

X

9.

¬[(G ⊃ H) ∨ (H ⊃ G)]✔

¬(G ⊃ H)✔

¬(H ⊃ G)✔

G

¬H

H

¬G

X

11.

¬(A ⊃ B) ⊃ C✔

¬C ∧ ¬B✔

¬¬A✔

A

¬C

¬B

¬¬(A ⊃ B)✔   C

A ⊃ B   X

¬A   B

X   X

17.

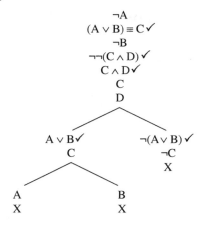

6.7.4

1.

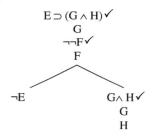

The proof begins by assuming the negation of the conclusion (the negation of ¬F) and so, in the tree, the conclusion is false. Since neither branch closes, there are two assignments of values that make the premises true while the conclusion is false. On the left branch, the conclusion is false and the premises true when G is true and E is false (and H is either true or false). On the right branch, the conclusion is false and the premises are true when H and G are true (and E is either true or false).

3.

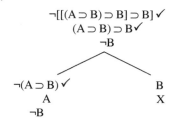

In this case there are no premises. The tree provides an interpretation that makes the assertion false. The left branch does not close and so the assertion is false when B is false (i.e., ¬B is true) and A is true.

5.

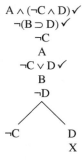

The left branch does not close and so the conclusion (C) is false when D is false, B is true, and A is true.

6.7.5.

1. This entailment does not hold. The entailment p $\Rightarrow$ p | ¬p does hold. If we knew that ¬p = q, then we could hold that p $\Rightarrow$ p | q. Since we don't have any information about q, the entailment does not hold.

3. Since $\square$(p $\wedge$ q) means that whenever p occurs then q occurs, it is not possible that there is a case in which p occurs and q doesn't, that is, ¬$\lozenge$(p $\wedge$ ¬q).

5. ¬$\lozenge$(S$\wedge$¬M) is equivalent to S | ¬ M. Given the Elimination rule, R | M we can then conclude R | S. This is equivalent to $\square$¬(R$\wedge$S). Notice that $\square$(R$\wedge$S) means that it is necessary that both R and S while $\square$¬(R$\wedge$S) means that not R and S is necessary. Therefore, $\square$(R$\wedge$S) does not hold.

7. The assertion on the left side of the entailment says that p holds and, if p holds, then ¬p does not hold. Since p $\supset$ ¬q is equivalent to ¬p $\vee$ ¬q and ¬p doesn't hold, then ¬q must hold. So the assertion on the left could be read '¬q $\wedge$ p'. So why can't we conclude that since p and ¬ q, p | q? The fact that ¬q and p are compatible (can both hold) does not show that p and q are incompatible. ¬q may have no relation whatever to p and the fact that they both hold at some point does not guarantee that p $\wedge$ q at some other point. p | q is a relation of necessity, but p $\supset$ ¬q is a hypothetical relation.

CHAPTER SEVEN: AN OVERVIEW OF QUANTIFIED LOGIC

7.4.

1. No managers are sympathetic. (Mx, Sx)
$\forall$x(Mx $\supset$ ¬Sx)

3. Some cell phones have no service here. (Cx, Sx)
$\exists$x(Cx $\wedge$ ¬Sx)

5. *Radiohead* concerts are amazing. (Rx, Ax)
$\forall$x(Rx $\supset$ Ax)

7. Not every earthquake is destructive. (Ex, Dx)
$\exists$x(Ex$\wedge$ ¬Dx)

9. Only registered voters can vote in the next election. (Rx, Vx)
   ∀x(Vx ⊃ Rx)

11. Grades are not always accurate. (Gx, Ax)
    ∃x(Gx∧ ¬Ax) or ¬∀x(Gx ⊃ Ax)

13. A person can be sued for libel if and only if they have written something against someone's character. (Px, Sx, Wx)
    ∀x[Px ⊃ (Sx ≡ Wx)]

15. Some hybrid cars are bad for the environment. (Hx, Bx)
    ∃x(Hx∧ Bx)

17. Most people feel more could have been done to prevent the genocide in Rwanda. (Px, Fx)
    ∃x(Px ∧ Fx)

19. If a soldier is not asked then that soldier is not required to tell. (Sx, Ax, Tx)
    ∀x[(Sx ∧ ¬Ax) ⊃ ¬Tx]

7.6.

1. All carpenters are precise and some handypersons are carpenters. Therefore, some handypersons are precise. (C, P, H)

    1. ∀x(Cx ⊃ Px) ∧ ∃x(Hx ∧ Cx)   ∴ ∃x(Hx ∧ Px)
    2. ∃x(Hx ∧ Cx)   1 Si
    3. Ha ∧ Ca   2 EI
    4. ∀x(Cx ⊃ Px)   3 Si
    5. Ca ⊃ Pa   4 UI
    6. Ca   3 Si
    7. Pa   5, 6 MP
    8. Ha   3 Si
    9. Ha ∧ Pa   8, 7 Con
    10. ∃x(Hx ∧ Px)   9 EG

3. All activists are loyal to a cause and all people in power conserve the status quo. Since no one who is loyal to a cause will conserve the status quo, it is not the case that some activists are people in power. (A, L, P, C)

    1. ∀x(Ax ⊃ Lx) ∧ ∀x(Px ⊃ Cx)
    2. ∀x(Lx ⊃ ¬Cx)   ∴ ¬∃x(Ax ∧ Px)
        3. ¬¬∃x(Ax ∧ Px)   Assume
        4. ∃x(Ax ∧ Px)   3 DN
        5. Aa ∧ Pa   4 EI
        6. La ⊃ ¬Ca   3 UI
        7. ∀x(Ax ⊃ Lx)   1 Si
        8. Aa ⊃ La   5 UI

9. $\forall x(Px \supset Cx)$	1 Si
10. $Pa \supset Ca$	6 UI
11. Aa	5 Si
12. La	8, 11 MP
13. Pa	5 Si
14. Ca	10, 13 MP
15. $\neg Ca$	6, 12 MP
16. $\neg Ca \wedge Ca$	14, 15 Con
17. $\neg \exists x(Ax \wedge Px)$	3-16 RA

5. If all store supervisors are wise, then some employees benefit. If there are some store supervisors who are not wise, then some employees benefit. As you can see, either way, some employees benefit. (S, W, E, B)

1. $\forall x(Sx \supset Wx) \supset \exists xEx$		
2. $\exists x(Sx \wedge \neg Wx) \supset \exists xEx$		$\therefore \exists xEx$
3. $\neg \exists xEx$	Assume	
4. $\neg \forall x(Sx \supset Wx)$	1, 3 MT	
5. $\neg \exists x(Sx \wedge \neg Wx)$	2, 3 MT	
6. $\exists x \neg(Sx \supset Wx)$	4 QE	
7. $\neg(Sa \supset Wa)$	6 EI	
8. $\neg(\neg Sa \vee Wa)$	7 DI	
9. $\neg \neg Sa \wedge \neg Wa$	8 DM	
10. $\forall x \neg(Sx \wedge \neg Wx)$	4 QE	
11. $\neg(Sa \wedge \neg Wa)$	10 UI	
12. $\neg Sa \vee \neg \neg Wa$	11 DM	
13. $\neg \neg Sa$	9 Si	
14. $\neg \neg Wa$	12, 13 DA	
15. $\neg Wa$	9 Si	
16. $\neg \neg Wa \wedge \neg Wa$	14, 15 Con	
17. $\exists xEx$	3-16 RA	

7. Everyone is a Democrat or a Republican, but not both. If someone is a Democrat, then she is a liberal or a conservative. All conservatives are Republican. So all Democrats are liberal. (D, R, L, C)

1. $\forall x[(Dx \vee Rx) \wedge \neg(Dx \wedge Rx)]$	
2. $\forall x[Dx \supset (Lx \vee Cx)]$	
3. $\forall x(Cx \supset Rx)$	$\therefore \forall x(Dx \supset Lx)$
4. Da	Assume
5. $(Da \vee Ra) \wedge \neg(Da \wedge Ra)$	1 UI

6. Da ⊃ (La ∨ Ca)	2 UI
7. La ∨ Ca	4, 6 MP
8. Ca ⊃ Ra	3 UI
9. ¬(Da ∧ Ra)	5 Si
10. ¬Da ∨ ¬Ra	9 DM
11. ¬¬Da	4 DN
12. ¬Ra	10, 11 DA
13. ¬Ca	8, 12 MT
14. La	7, 13 DA
15. Da ⊃ La	4-14 CA
16. ∀x(Dx ⊃ Lx)	15 UG

9. If everyone is a liberal, then no one is conservative. There is a governor of Alaska and she is a conservative. So at least someone is not a liberal. (L, C, G)

1. ∀xLx ⊃ ¬∃xCx	
2. ∃x(Gx ∧ Cx)	∴ ∃x¬Lx
3. ¬∃x¬Lx	Assume
4. ∀x¬¬Lx	3 QE
5. ∀xLx	4 DN
6. ¬∃xCx	1, 5 MP
7. ∀x¬Cx	6 QE
8. Ga ∧ Ca	2 EI
9. ¬Ca	7 UI
10. Ca	8 Si
11. ¬Ca ∧ Ca	9, 10 Con
12. ∃x¬Lx	3-11 RA

7.9.2.

1.

$$\forall x(Ax \supset \neg Bx)$$
$$\neg\neg\exists x(Ax \land Bx)\checkmark$$
$$\exists x(Ax \land Bx)\checkmark$$
$$Aa \land Ba\checkmark$$
$$Aa$$
$$Ba$$
$$Aa \supset \neg Ba\checkmark$$

```
        ╱‾‾‾‾‾‾‾╲
      ¬Aa          ¬Ba
       X            X
```

3.

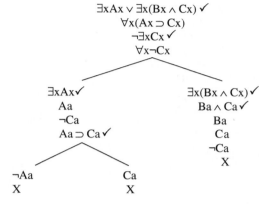

∃xAx ∨ ∃x(Bx ∧ Cx) ✓
∀x(Ax ⊃ Cx)
¬∃xCx ✓
∀x¬Cx

∃xAx✓
Aa
¬Ca
Aa ⊃ Ca ✓

¬Aa
X

Ca
X

∃x(Bx ∧ Cx)✓
Ba ∧ Ca✓
Ba
Ca
¬Ca
X

5.

∀x(¬Fx ⊃ Gx)
¬∃xFx✓
¬∀xGx✓
∀x¬Fx
∃x¬Gx✓
¬Ga
¬Fa
¬Fa ⊃ Ga✓

¬¬Fa
X

Ga
X

7.9.3

1.

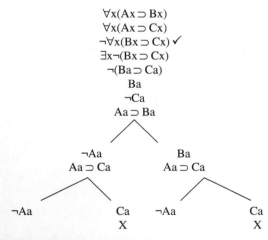

∀x(Ax ⊃ Bx)
∀x(Ax ⊃ Cx)
¬∀x(Bx ⊃ Cx) ✓
∃x¬(Bx ⊃ Cx)
¬(Ba ⊃ Ca)
Ba
¬Ca
Aa ⊃ Ba

¬Aa
Aa ⊃ Ca

¬Aa

Ca
X

Ba
Aa ⊃ Ca

¬Aa

Ca
X

Two branches do not close. The premises are true and the conclusion is false
when there is a something, a, that is not A, is B, and is not C.

3.

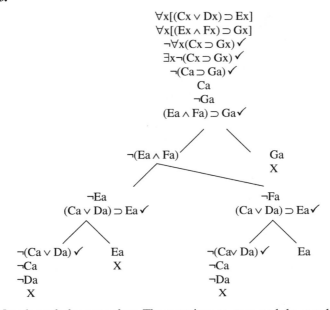

One branch does not close. The premises are true and the conclusion false when there is a thing, a, such that it is E and C and not F or G.

Bibliography

Addams, Jane. 1915. The Revolt Against War. *The Survey*, 34 (July 17), 355–359.

American Anthropological Association. 1998. *Statement on "Race"* (May 17, 1998). http://www.aaanet.org/stmts/racepp.htm (accessed January 5, 2008).

Saint Anselm. 1903. *Proslogium; Monologium: An Appendix in Behalf of the Fool by Gaunilo; And Cur Deus Homo*, Translated by Sidney Norton Deane. Chicago: The Open Court Publishing Company.

Anzaldúa, Gloria. 1999. *Borderlands/La Frontera: The New Mestiza*, Second Edition. San Francisco: Aunt Lute Books.

Saint Thomas Aquinas. 1962, *Aristotle: On Interpretation. Commentary by Saint Thomas and Cajetan*. Translated by Jean T. Oesterle. Milwaukee: Marquette University Press.

Aristotle. 1941. *The Basic Works of Aristotle*. Edited by Richard McKeon. New York: Random House.

Barad, Karen. 2007. *Meeting the Universe Halfway: Quantum Physics and the Entanglement of Matter and Meaning*. Durham and London: Duke University Press.

Berkeley. George. 1776. *Three Dialogues Between Hylas and Philonous*. London.

Bosanquet, Bernard. 1911. *Logic or the Morphology of Knowledge*, Two Volumes. Bristol: Thoemmes Press, 1999.

Bunge, Robert. 1984. *An American Urphilosophie: An American Philosophy, BP (Before Pragmatism)*. Lanham, MD: University Press of America.

Bush, George W. 2001. President Unveils Back to Work Relief Package. http://www.whitehouse.gov/news/releases/2001/10/20011004-8.html, accessed March 4, 2005.

Bush, George W. 2002. President Bush Outlines Iraqi Threat. http://www.whitehouse.gov/news/releases/2002/10/20021007-8.html, accessed January 5, 2009.

Cajete, Gregory. 2000. *Native Science*. Albuquerque, NM: Clear Light.

Carnap, Rudolf. 1959. The Elimination of Metaphysics through the Logical Analysis of Language. In *Logical Positivism*. Edited by A. J. Ayer. New York: Free Press, 60–81.

Carnap, Rudolf. 1967. *The Logical Structure of the World & Pseudoproblems in Philosophy*. Translated by Rolf A. George. Berkeley: University of California Press.

Cohen, Morris R. 1944. *A Preface to Logic*. New York: Henry Holt and Company.

Cohen, Morris R., and Ernest Nagel. 1934. *An Introduction to Logic and Scientific Method*. New York: Harcourt, Brace & World, Inc.

Logic: Inquiry, Argument, and Order, by Scott L. Pratt
Copyright © 2010 John Wiley & Sons, Inc.

Davidson, Donald. 2001. *Inquiries into Truth and Interpretation*, Second Edition. Oxford: Clarendon Press.

Deloria, Vine, Jr. and Daniel R. Wildcat. 2001. *Power and Place: Indian Education in America*. Golden, CO: Fulcrum Publishing.

Deloria, Jr., Vine. 1969. *Custer Died for Your Sins*. Norman, OK: University of Oklahoma Press, 1988.

Deloria, Jr., Vine. 1999. *Spirit and Reason: The Vine Deloria, Jr. Reader*. Golden, CO: Fulcrum Publishing.

Descartes, Rene. 1911. *The Philosophical Works of Descartes*. Translated by Elizabeth S. Haldane. Cambridge: Cambridge University Press.

Dewey, John. 1925. *Experience and Nature*. In *Later Works*, Volume 1. Edited by Jo Ann Boydston. Carbondale, IL: Southern Illinois University Press, 1981.

Dewey, John. 1933. *How We Think*, Revised edition. In *Later Works*, Volume 8. Edited by Jo Ann Boydston. Carbondale, IL: Southern Illinois University Press, 1986.

Dewey, John. 1937. Review of *The Collected Papers of C. S. Peirce*, 6 volumes. In *Later Works*, Volume 11. Edited by Jo Ann Boydston. Carbondale, IL: Southern Illinois University Press, 1989.

Dewey, John. 1938. *Logic: The Theory of Inquiry*. In *Later Works*, Volume 12. Edited by Jo Ann Boydston. Carbondale, IL: Southern Illinois University Press, 1986.

Dewey, John. 1938b. The Unity of Science as a Social Problem. In *Later Works*, Volume 13. Edited by Jo Ann Boydston. Carbondale: Southern Illinois University Press, 1988.

Du Bois, W. E. B. 1897. The Conservation of Races. In *The American Negro Academy Occasional Papers*, No. 2.

Du Bois, W. E. B. 1903. *The Souls of Black Folk*. Chicago: A. C. McClurg and Co.

Du Bois, W. E. B. 1920. *Darkwater: Voices from Within the Veil*. New York: Harcourt, Brace, and Howe.

Dussel, Enrique. 1988. *Ethics and Community*. Eugene, OR: Wipf & Stock.

Edwards, Jonathan. 1749. *The Life of the Late Reverend David Brainerd*. Boston: D. Henchman.

Fanon, Franz. 1963. *The Wretched of the Earth*. Translated by Constance Farrington. New York: Grove Press, 1968.

Field, Hartry. 1994. Deflationist Views of Meaning and Content. *Mind*, 103, 411, 249–85.

Follett, Mary Parker. 1924. *Creative Experience*. New York: Longmans, Green, and Co.

Frege, Gottlob. 1918. The Thought. Translated by A. M. and Marcelle Quinton. *Mind*, LXV, July 1956, 289–311.

Gilligan, Carol. 1993. *In a Different Voice: Psychological Theory and Women's Psychological Development*. Cambridge: Harvard University Press.

Goldman, Emma. 1911. *Anarchism: What It Really Stands For*. New York: Mother Earth Publishing Association.

Goodman, Nelson. 1983. *Fact, Fiction and Forecast*, Fourth Edition. Cambridge: Harvard University Press.

Habermas, Jürgen. 1971. *Knowledge and Human Interests*. Translated by Jeremy J. Shapiro. Boston: Beacon Press.

Habermas, Jürgen. 1984. *Theory of Communicative Action*, Volume 1. Translated by Thomas McCarthy. Boston: Beacon Press.

Habermas, Jürgen. 1998. Actions, Speech Acts, Linguistically Mediated Interactions, and the Lifeworld. In *On the Pragmatics of Communication*. Edited by Maeve Cooke. Cambridge: MIT Press, 215–256.

Harris, Leonard, Scott L. Pratt, & Anne S. Waters, eds. 2002. *American Philosophies: An Anthology*. Malden, MA: Blackwell Publishers.

Hegel, Georg W. F. 1975. *Hegel's Logic*. Translated by William Wallace. Oxford: Clarendon Press.

Hewitt, J. N. B. 1902. Orenda and a Definition of Religion. *American Anthropologist*, 4, 33–46.

Hume, David. 1779. *Dialogues Concerning Natural Religion*, Second Edition. London.

Johnson, Mark. 2007. *The Meaning of the Body: Aesthetics of Human Understanding*. Chicago: University of Chicago Press.

Jones, William. 1905. The Algonkin Manitou. *Journal of American Folklore*, 18:70, 183–90.

Kant, Immanuel. 1974. *Logic*. Translated by Robert S. Hartman and Wolfgang Schwartz, New York: Dover.

Kuhn, Thomas. 1970. *The Structure of Scientific Revolutions*, Second Edition, Enlarged. Chicago: University of Chicago Press.

Langer, Susanne. 1967. *An Introduction to Symbolic Logic*, Third Revised Edition. New York: Dover Publications.

Leibniz, Gottfried. 1765. *New Essays on Human Understanding*. Edited and translated by Peter Remnant and Jonathan Bennett. Cambridge: Cambridge University Press.

Lewis, C. I. 1918. *A Survey of Symbolic Logic*, Berkeley: University of California Press.

Lewis, C. I. and C. H. Langford. 1959. *Symbolic Logic*, Second Edition, New York: Dover Publications, Inc.

Lovejoy, Arthur O. 1936. *The Great Chain of Being: A Study of the History of an Idea*. New York: Harper & Row.

Lucretius. 1966. On the Nature of Things. In *Greek and Roman Philosophy after Aristotle*. Edited by Jason L. Saunders. New York: The Free Press, 15–46.

Lugones, Maria. 2007. Homosexualism and Colonial/Modern Gender. *Hypatia*, 22(1), 186–209.

Lynch, Michael P. 2004. *True to Life: Why Truth Matters*. Cambridge, MA: MIT Press.

Mather, Cotton. 1853. *Magnalia Christi Americana: The Ecclesiastical History of New England*. Hartford, CT.

Mendieta, Eduardo. 2000. "The Making of New Peoples," in *Hispanics/Latinos in the United States*. Edited by Jorge J. E. Gracia and Pablo De Greiff. New York: Routledge.

Mill, John Stuart. 1863. *On Liberty*. Boston: Tichnor and Fields.

Mill, John Stuart. 1875. *A System of Logic*, two volumes. London: Longmans, Green, Reader, and Dyer.

Mill, John Stuart. 1924. *Autobiography*. New York: Columbia University Press.

Mohanty, Chandra Talpade. 2004. *Feminism Without Borders: Decolonizing Theory, Practicing Solidarity*. Durham, NC: Duke University Press.

Nasr, Seyyed Hossein. 2003. Reflections on Islam and Modern Life, in *Al-Serat: A Journal of Islamic Studies*. http://www.al-islam.org/al-serat/toc.htm, accessed July 25, 2003.

Nye, Andrea. 1990. *Words of Power: A Feminist Reading of the History of Logic*. New York & London: Routledge.

Ockham, William. 1964. William Ockham on Epistemological Problems. In *Medieval Philosophy: Selected Readings from Augustine to Buridan*. Edited and introduced by Herman Shapiro. New York: Modern Library, 482–508.

Peirce, C. S. 1994. *The Collected Papers of Charles Sanders Peirce*, Electronic Edition, with an editorial introduction by John Deely, reproducing reproducing Vols. I–VI ed. Charles Hartshorne and Paul Weiss (Cambridge, MA: Harvard University Press, 1931–1935), Vols. VII–VIII ed. Arthur W. Burks (same publisher, 1958).

Plato. 1892. *The Dialogues of Plato*, Volume III, Third Edition. Translated by B. Jowett. Oxford: Clarendon Press.

Plato. 1909. *The Apology, Phaedo and Crito*. Translated by B. Jowett. Edited by C. W. Eliot. New York: P. F. Collier and Son.

Plato. 1961. *The Collected Dialogues of Plato*. Edited by Edith Hamilton and Huntington Cairns. Princeton: Princeton University Press.

Pratt, Scott L. 2006. Persons in Place: The Agent Ontology of Vine Deloria, Jr. *APA Newsletter on American Indians in Philosophy*, 6(1), 4–9.

Pratt, Scott L. 2007. The Experience of Pluralism. *Journal of Speculative Philosophy*, 21(2), 106–14.

Plumwood, Val. 2002. The Politics of Reason: Toward a Feminist Logic, in *Representing Reason: Feminist Theory and Formal Logic*. Edited by Rachel Joffe Falmagne and Marjorie Hass. Lanham, Boulder, New York, Oxford: Rowman and Littlefield.

Priest, Graham. 2006. *In Contradiction*, Second Edition. Oxford: Clarendon Press.

Priest, Graham. 2008. *An Introduction to Non-Classical Logic*, Second Edition. Cambridge: Cambridge University Press.

Quine, W. V. O. 1965. *Elementary Logic*, revised ed. New York: Harper and Row.

Quine, W. V. O. 1980. Two Dogmas of Empiricism. In *From a Logical Point of View*. Cambridge: Harvard University Press.

Randall, John Herman, Jr. 1963. *How Philosophy Uses Its Past*. New York: Columbia University Press.

Robinson, Daniel Sommer. 1947. *The Principles of Reasoning : An Introduction to Logic and Scientific Method*, Third Edition. New York: Appleton-Century-Crofts.

Royce, Josiah. 1896. *The Spirit of Modern Philosophy*. Boston: Houghton, Mifflin and Company.

Royce, Josiah. 1899. *The World and the Individual First Series*. New York: Dover, 1959.

Royce, Josiah. 1913. Order. In *Royce's Logical Essays: Collected Logical Essays of Josiah Royce*. Edited by Daniel S. Robinson. Dubuque, IA: Wm. C. Brown Company, 1951, 204–231.

Royce, Josiah. 1914. *The Principles of Logic*. In *Royce's Logical Essays: Collected Logical Essays of Josiah Royce*. Edited by Daniel S. Robinson. Dubuque, IA: Wm. C. Brown Company, 1951, 310–378.

Royce, Josiah. 1919. *Lectures in Modern Idealism*. New Haven: Yale University Press, 1964.

Russell, Bertrand. 1938. *The Principles of Mathematics*, Second Ed. New York & London: W. W. Norton & Company, 1996.

Russell, Bertrand. 1950. Philosophy for Laymen. In *Unpopular Essays*. New York: Simon and Schuster.

Russell, Bertrand. 1962. *An Inquiry into Meaning and Truth*. Middlesex and Baltimore: Pelican Books.

Sagoyewatha (Red Jacket). 1866. In *The Life and Times of Sa-Go-Ye-Wat-Ha, or Red Jacket*, by William L. Stone. Albany, NY: J. Munsell, 288–91.

Schopenhauer, Arthur. 1908. *Studies in Pessimism*. Translated by T. B. Saunders. New York: Macmillan.

Schutte, Ofelia. 1993. *Cultural Identity and Social Liberation in Latin American Thought*. Albany, NY: State University of New York Press.

Schutte, Ofelia. 2000. Cultural Alterity: Cross-Cultural Communication and Feminist Theory in North-South Contexts. In *Decentering the Center: Philosophy for a Multicultural, Postcolonial, and Feminist World*. Edited by Uma Narayan and Sandra Harding. Bloomington, IN: Indiana University Press.

Sharp, Henry S. 1995. Asymmetric Equals: Women and Men about the Chipewyan. In *Women and Power in Native North America*. Edited by Klein and Ackerman. Norman, OK.: University of Oklahoma Press.

Smith, Adam. 1817. *Wealth of Nations*. London: Doig and Stirling.

Thurman, Howard. 1965. *The Luminous Darkness*. New York: Harper & Row.

Turing, Alan. 1950. Computing machinery and intelligence. *Mind*, 59, 433–460.

Wallerstein, Immanuel. 2004. *World-System Analysis: An Introduction*. Durham, NC: Duke University Press.

Whitney, Stephen. 1985. *Western Forests*. New York: Alred A. Knopf.

Whitt, Laurie Anne. 1998. Cultural Imperialism and the Marketing of Native America. In *Natives and Academics: Research and Writing about American Indians*. Edited by Devon A. Mihesuah. Lincoln, NE: University of Nebraska Press, 139–71.

Wollstonecraft, Mary. 1891. *A Vindication of the Rights of Women*. London: T. Fisher Unwin.

Zack, Naomi. 1993. *Race and Mixed Race*. Philadelphia: Temple University Press.

Index

Abduction, 20, 47, 48, 54, 61, 78, 95–96, 129, 165–166, 219n3. *See also* Forming hypotheses; Suggestion
  and nominalism and realism, 92–93
  defined, 43–46
  principles of, 46
Ad baculum (fallacy), 82, 83
Ad hominem (fallacy), 81–82
Ad misericordium (fallacy), 82, 222
Ad populum (fallacy), 82–83
Abstraction, the problem of, 2–5, 220, 222
Addams, Jane, 65
Addition (argument), 188, 210
Affirmative assertions (propositions), 100–101, 101n5, 113–116, 211–214
Affirming the consequent (fallacy), 84–85, 177
Agent (agency), 10, 63, 73, 92, 93, 95, 218n2
  defined, 125, 170
  and judgment, 174, 175–176
  ontology, 169–171, 173, 199–202
  problems of logic, 218–222, 219n3
Agreement, method of, 146, 147
Ambiguity (fallacy), 79
American Anthropological Association, 18–19
Amphiboly (fallacy), 79
Ampliative reasoning, 129, 139, 165–166
Analogical arguments, 20–21, 135, 139–142
  standards, 140–142
Anselm, Saint, 65–66
Antecedent of a conditional, 39, 184
Anzaldúa, Gloria, 13, 219–220, 221
Appeal to authority (fallacy), 80
Aquinas, Thomas, 16, 97

Argument, 19–23, 35, 46–47, 61, 165–166, 197–198, 222. *See also* Validity; Fallacies
  Aristotle on, 97–98
  defined, 20, 99
  and fallacies, 75–76, 77–78
  inductive, 129, 130–132, 134–135, 139–140, 145–146, 151
  as inquiry, 47, 52–57
  valid forms, 39, 72–75, 96, 99, 112–114, 120, 142, 177, 178, 185–186, 190, 193
  warrant, 91–92
Argumentum ad ignoratiam (fallacy), 83
Aristotle, continuity, 7, 14, 22–23, 83
  on fallacies, 75–76
  theory of the syllogism, 97–99, 100, 103, 112–113, 119, 126, 127, 128n3, 128–129, 166
Assertion(s), *see also* Proposition(s); Truth
  defined, 21, 62, 93, 99, 169
  functions of, 94–96
  logic of, 178–183
  and speech acts, 72–75, 168
  standard form, 98–104
  tableaux, 190–194
  and truth, 36–39, 43, 45
  limits of, 198–199
Association (rule), 180
Axioms of quality, 113
Axioms of quantity, 113
Axiomatic proofs of validity, 32, 113–116, 168, 178, 198

Background knowledge, 68, 69–71
Barad, Karen, 169, 218–219
Bayes' theorem, 157–159, 160
Belonging, 173
Berkeley, George, 67–68

Logic: Inquiry, Argument, and Order, by Scott L. Pratt
Copyright © 2010 John Wiley & Sons, Inc.